木の箱

広橋桂子 KEIKO HIROHASHI
杉村敏男 TOSHIO SUGIMURA
U.G.サトー U.G. SATO
岡田宏三 KOZO OKADA
新井 亮 RYO ARAI
新井明子 AKIKO ARAI
秋月 繁 SHIGERU AKIZUKI
撮影● 福田匡伸 Photographer MASANOBU FUKUDA

六耀社

Wood Package 木の箱

目次

CONTENTS

箱をつくる木 木でつくる箱

横山正（東京大学助教授）

1.
A folk craft box of Indios in Mexico. it is a kind of souvenir for tourists. I am captivated by the unfailing force in its pattern, and asked someone visiting the country to buy some boxes for me. Thus, I have quite a number of boxes of similar type, but the pair of boxes I obtained for the first time seem to be the best. They are made of a kind of pine tree with its surface slightly engraved and colored.

2.
The blue lacquered box on the left is from Hsian, the ancient capital of China. China can be said as a country of boxes. Wherever you go, you'll find various, kinds of boxes. But there are few sophisticated ones like this hexahedral box. The inside is lacquered in black and gold like pear's skin. The right one is a box by a contemporary American designer, whose name I cannot recall to my regret. He inlaid plastics picking up a design idea from Indian patterns.

3.
A folk craft souvenir from Peru. A friend of mine found it in a shop in Harajuku, Tokyo.
The same flower is drawn on both outside and inside of the box. On the back of the swelling lid, a crowd of dolls in festive dresses are standing. It is quite dramatic. There is no other such fantastic one in my collection. As I look at the face of each doll, I feel light-hearted.

人間はいちばん最初、何で箱を造っただろうか。やきものとか金属ではなさそうだが、とするといったい何だろうか。「日本美術工芸」に京都国立博物館の灰野昭郎氏が「漆・箱」という日本の漆塗りの箱の歴史についての興味深い論文を連載されたことがあるが、そのなかで氏は、「日本の箱の嚆矢は皮であったといいたい」と言っておられる。といっても根拠あってのことではなくて、まずは奈良時代の箱の主流がそうであったからということのようだ。しかし木で箱をつくるとか、竹で箱をつくるというのは、いちおうの工作技術をともなうことで、たしかにその点から考えると、やはりその出現は少し遅れてのことであったろうという気がする。これに対して皮のほうは、正倉院の漆皮の箱のような立派なものはべつとして、ごく簡単な容れものを造るということなら糸と針で縫い合わせれば良いのだから、たしかに発生的にはこちらのほうが古い可能性がある、というより、そう考えてまず間違いはないのだろう。(ただ正倉院の漆皮箱のようなものの工法についてはそんな単純な話はできないようだ。これはあらかじめ造った木型の上に一枚の皮を被せ、それを打って延ばしながら木型になじませ、変形しないように漆で固めるというもので、この木型はあとではずしやすいよう、複数のパーツを枘を用いて組立てたものが用いられていたようだから、つまりは木製の箱の製作技術に類するものがあってはじめて可能な造りということになる。)

革に代わって木の箱が主流になっていくのは、やはりまずはその堅固さゆえのことに違いない。内容物を護る

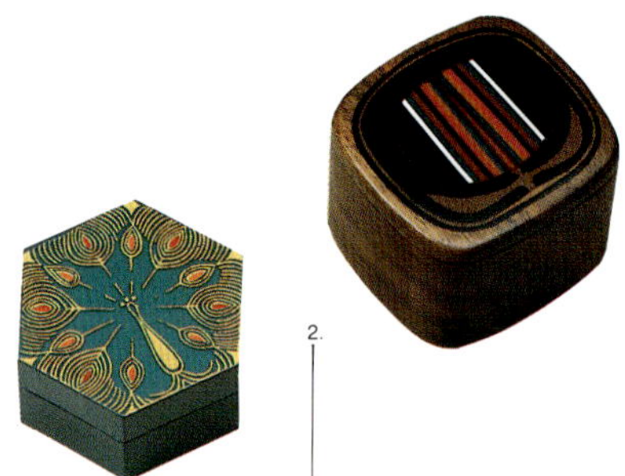

2.

1.

3.

メキシコのインディオの民芸品の箱。現代の観光お土産用のものだが、本来の文様の力強さがおちていず、感心させられる。私はこれに魅かれてメキシコ旅行をするひとに頼んだりしてこの手のものをいっぱい蒐めたが、やはり最初に求めたこの二つが結局はいちばん良いようだ。松のたぐいの木の表面を浅く彫り込んで彩色している。
大W×D×H：255×140×110㎜
小185×100×85㎜

左の青い漆塗りの箱は中国の古都、西安のもの。中国は箱の国と言ってもいいくらい、どこに行っても、興味深い箱を発見する。でもこの六角の箱くらい洗練されたものは珍しいだろう。内側は黒地に金の梨地である。
右はアメリカの現代のデザイナーの作品。インディアンの文様をヒントに、プラスティックを象嵌したもの。どうしてもデザイナーの名前が思い出せないのが残念である。
中国のもの 64×64×33㎜
アメリカのもの 75×70×48㎜

ペルーの民芸品のお土産。といってもこれは知人が原宿のお店屋さんで見つけてくれたもの。内側にも、そのまま外と同じ描きかたで花が描かれていて、おまけにふくらんだ蓋の裏にお祭りらしい人形の群れがくっついているのが大傑作である。ぼくのコレクションの中にも、こんな奇想天外なものはない。じっと一人一人の表情を眺めていると楽しくて、うきうきしてくる。
90×68×70㎜

4.
A cash box of 17-century Spain. On the two sides of a house-shaped box are silver colored horns, somehow, bringing up the image of bullfighting. Besides, the inside is colored in blazing red!
I happened to buy this one at the words of the owner of an antique shop who said that a famous dyeing artist Keisuke Serizawa was going to buy it before long.
And I admit this was not a bad buy. Whenever I see this box, I think of him.

5.
I bought this at a morning market in Kyoto, but I don't know where it was made, nor do I have the hope of tracing it. Shells are inlaid in black lacquered surface.
The inside is painted in flaming red. It has a mysterious beauty which attracts my attention. I wonder if this came from South China or somewhere from Southeast Asia. Even when the lid is closed, the red inside shows slightly along the joint.

5.

この箱は京都の朝市で求めたものだが、まったく産地不明。ほとんどいま出来なのではないかと思うのだが、どこのものか分らない。黒の漆のなかに貝殻が象嵌されて内側は真紅。どこか妖しいような美しささえあって、何となく気にかかる箱である。中国の南のほうか東南アジアのどこかのものか。蓋を閉めても、合わせ口のところにすっと内側の紅の色がのぞく。
104×75×37㎜

という目的からは、周囲が堅固であるにこしたことはないからである。それに皮ならばゆがみや亀裂といったこともあるだろう。だから堅固ということに加えて、かちっとした形を保持するということも木の箱の魅力になったはずである。またさきに引いた灰野氏が、装飾を重んじられるようになったことが木の箱を求めさせた理由のひとつではないかと推察しておられるのも興味深い。思うに箱には元来、二つのありかたがある。表に出てたえず目に触れるところに置かれるものと、たんなる収納具として納戸や蔵にしまわれるものとである。前者は言うまでもなく華やかな存在であり、蒔絵を施されたり彫り込まれたりと、いろいろに装飾されるはずだし、後者は仮にお金をかける場合でも地味に造られ、ほとんど装飾は施されない。そしてもし箱が表立った存在になったとするならば、やはり装飾の可能性が大きいものが求められるであろう。皮の箱にしても、李朝や中国のそれに見るように、截金を貼るなどいろいろ装飾を施した美しいものを見るけれども、やはり木の箱の表現の自由度にはかなわない。金属の箱ややきものの箱というのもあるけれども、何と言っても表現の幅広く多彩なのは木の箱であろう。ここに少しお目にかけたように、私はややがらくためいた箱をあれこれ蒐めているのだけれども、そのささやかなコレクションのなかでさえも木の箱の表現のヴァラエティはおどろくほどである。

といってもこの本でデザイナーたちが手掛けている箱はむしろ庶民の箱の伝統につながるもので、豪華な蒔絵の箱などとは別の世界であり、つまりは私のコレクションの内容にも通じるものなのだが、それでもここに見る

4.

スペインの17世紀の銭箱。家みたいな形をしたものの棟の両側に銀色の角が出ていて、何となく闘牛を想像させる。それにこれは蓋を開けると中が真赤に塗られているのである!ぼくはこれを芹沢銈介さんが買いたいと言っていらっしゃいましたという「無尽蔵」の親父の口車に乗って買ってしまったのだが、でももの自体はそんなに悪くない。
235×115×150㎜

6.
The boxes covered or wedged with metal excite with talismanical interest. The combination of wood and metal contains a sort of absurdity. As a result, a special charm is brought about which can never be felt with boxes made solely of wood.
The left one is a contemporary box from Vienna.
The upper one is a contemporary box from Florence.
The small one below is an contemporary box from India.
The right one may be a Korean box of the recent past.

通り、実にさまざまな表現をもっている。こういう技術は日本ではいったい、いつごろから開発されていったのだろうか。伊勢神宮の御神宝に有名な柳筥というのがある。断面を三角にした柳の白木の細棒を絹糸でつないで面をつくるものだが、これは板をつくるという技術以前の面の造りかたを伝えたものであろう。柳を使うのはそれがやはりある種の信仰につながる面をもったからで、椿の枝同様、柳の枝も呪術的な意味を持つと考えられたのであろう。つまり、この筥は原初的には柳の枝をかがって造られたに違いないのである。しかし奈良時代に入ると、正倉院にたくさん素晴らしい箱の実例が残っていて、もう枘を切って組合わせることもやられているし、釘も、また一種の接着剤も用いられていて、ほとんど木の箱の製作技術が完成していることが知られる。

けれども、これらはすべて上つ方のものである。庶民がこんな上等なものを日常用いられた訳はない。ただ鎌倉時代の絵巻類を見れば、庶民の生活のなかに木を組立てたものや曲げ物の形での容器がいろいろ登場しており、実用性をもったものとしての箱はすでに平安時代に庶民の生活にも登場していたはずである。ただ庶民は箱に入れなければならない「もの」をそんなに沢山は持っていなかったはずだし、また箱の製作は手数のかかるものであったから、身のまわりにあるその数はまことに僅かであったに違いない。これが大量に造られ庶民の生活のなかに定着するのはやはり江戸時代であって、それには室町時代から安土桃山時代にかけての木工具の技術革新がひとつの背景をなしていよう。

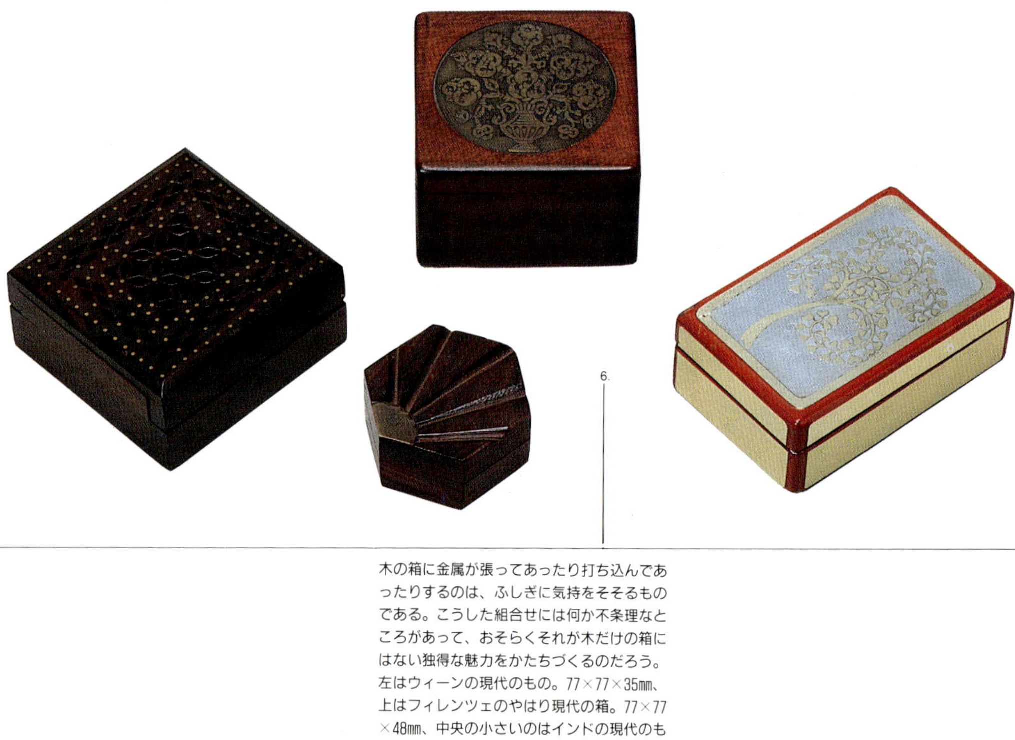

6.

木の箱に金属が張ってあったり打ち込んであったりするのは、ふしぎに気持をそそるものである。こうした組合せには何か不条理なところがあって、おそらくそれが木だけの箱にはない独得な魅力をかたちづくるのだろう。左はウィーンの現代のもの。77×77×35mm、上はフィレンツェのやはり現代の箱。77×77×48mm、中央の小さいのはインドの現代のもの。51×51×24mm、右は朝鮮の少し前のものか。98×63×35mm

7.
A wood-carved box from India. I often see souvenir boxes from India which do not fit my feelings. This is not an antique but I use it as I like the artless shape of the deer. This box seems to have been unvarnished when it was made, but painted in black color by someone at a later stage. In this stick-type box, a hinge runs through a hole bored on the side, so that it won't be seen from outside.

8.
A lacquered box of Punjab, India. The lacquer of India has too strong smell to put something inside when it is applied to a box. However, the carefree pattern is beautiful enough to catch viewers' hearts. Among my collection, this box with the picture of birds on a tree is the most extravagant, and I keep it very preciously. Bird lovers seem unable to contain their desire of possessing this, and in fact, one of them reserved this as my memento.

それぞれの目的に応じてのさまざまな箱が造られ、それが庶民の日常目に触れるところに氾濫していったさまは、江戸時代の浮世絵の描写に見ることが出来る。江戸の町人文化の爛熟は、また精巧な細工や名人芸を誇る方向へと導いていったが、もちろんそれは箱の世界へも投影された。上流階級の箱の文化はずっと早く、すでに鎌倉や室町といった時代に花開いたけれど、上層町人もふくめての一般の箱の文化が花開いたのは、やはり江戸時代と言ってよいのではあるまいか。

今回のデザイナーたちによる箱の製作を請負った春日部について、江戸時代の中期に箱細工をやっていたことを証す文献があるというのもそのひとつのあらわれである。春日部は日光街道にあり、東照宮の工事に携わった大工をはじめさまざまな職人と接触する機会も多く、そんなことから指物仕事が根をおろしたらしい。今回、埼玉県の工業技術研究所の影山和則氏からいろいろ春日部の歴史と現況についてお話を伺ったのだが、その中でこれまでの考えを改めなければならないことを知った。私は桐簞笥で有名な春日部は、まず簞笥の製造で大きくなり、その端材を使って箱を造ったのだと考えていたのだが、事実はどうも逆で、はじめ箱指物をやっていたのがしだいに大形のものを手掛けるようになったらしい。影山さんのお話では、浅草の問屋街に近いということも、そうした産業が栄えた理由のひとつだろうということである。何でも、大正から昭和のはじめというのが春日部の箱生産の最盛期で、歯磨の箱などを造ったのを大八車で得意先の浅草の問屋まで一日二往復して運んだそうで

7.

インドの木彫の箱。インドの今出来のお土産品の箱はしょっちゅうお目にかかるが、ちっとも気に入るのが無い。これは少し前というくらいで別に古いものでは無いが、素朴な鹿の彫りが気に入って物入れに愛用している。これはもともと生地仕上げだったのを、あとで誰かが黒く塗ったのだとおもう。この箱は棒状の蝶番が側板をくり抜いて通してあって、いっさい金物が見えないようになっている。
206×130×50㎜

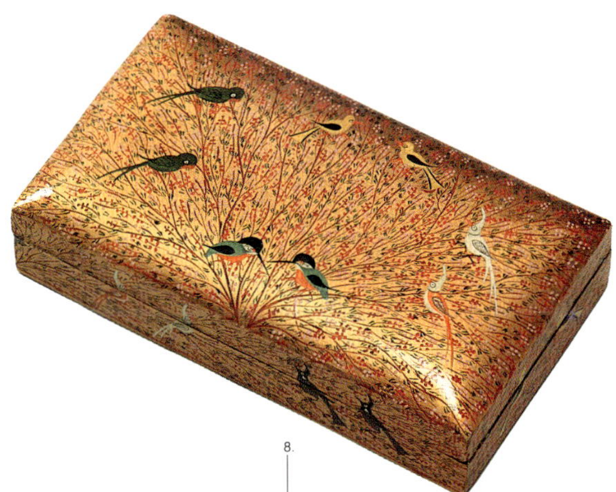

8.

インドのパンジャブの漆塗りの箱。インドの漆はとても臭くて中にものを入れる気がしないのが欠点だが、土産ものとは言え、その外側の文様は実にのびやかにまた美しく出来ていて人の心を魅く。いろいろ持っているなかでもこの木に鳥がとまっている柄のがいちばん豪奢で大事にしているのだが、鳥好きのひとにはたまらないらしくて、いつでも引取るという予約をしているひとが何人もいる。
177×99×40㎜

9.
Saving-boxes made in the recent past in France.
They have nothing special but have friendly appearance.
Though they look very simple, they are pieces of elaborate work. The larger one has inlays and the smaller one has sophisticated metal fittings. These are, of course, put on my desk, serving their original purpose.

10.
A tea box of the Victorian Period of Great Britain.
The inside of the box is separated into two parts, and under the lid with an ivory handle is a tea container with tin leaf.
This type of box is now in fashion in Japan and many of its kind are on the shelves of antique shops.
However, in comparison, the one I have seems by far better both in proportion and nobility. Is it because of my effusive attachment for it?
The Victorian tea box always locks.

ある。春日部が箱の産地となったのは、もちろん桐の木があったからで、昭和の初めまでは自前の桐を使っていたのだが、それもそうした盛況のなかで使い尽くされてしまったという。もちろんいまは秋田など日本の各地から、さらには北米、台湾から桐材を輸入して用いているという話である。

ところで春日部の名を高からしめた桐という木だが、以前これがさきほどの歯磨の箱などいろいろな用途にさかんに使われたのは、おそらく日本に木がたくさんあってとくに高くなく、それに軽くて（本邦産の木材のなかでもっとも軽い）手軽だったからであろう。しかしそれとは別に、骨董品のやきものや漆器は必ず桐箱におさめられている。べつにとくに堅牢とは言い難い、むしろ逆の材なのだが、それが用いられるのはむしろそのソフトな感じが大事な内容物を痛めないというイメージがあるからだろうか。しかし桐材の箱が好まれるのはどうもそれだけではなさそうで、木肌が美しいこと、軽いということももちろんあるが、それに例の桐の箱は火難、水難に強いという言い伝えが多分に桐を択ばせているのではないかという気がする。すなわち桐材は湿気を通さないので水害にあったが桐簞笥のなかの着物が濡れなかったとか、あるいは火事があったが桐の簞笥が燃えずに残ったとかいうたぐいの話である。桐材は吸湿性、吸水性が低いほうだが、しかしそれもほぼ同比重の他の木とくらべてそんなに大差はない。ただ全体としては低いほうなので、いくぶんその傾向はあるという程度であろうか。それに収縮率も小さい（これも比重によるので、ほぼ同比重の他の木にくらべれば大差ない）から、抽出などが

9.

10.

フランスのちょっと昔の貯金箱二つ。何の変哲もないものだけれど、人なつこいところがある。もっとも何でもないようでいて実は結構手が込んでいるので、大きいほうは木の象嵌がしてあるし、小さいほうの金具だってなかなか凝っているではないか。もちろん机上にあって、ちゃんと当初通りの役をつとめている。
大130×82×86mm
小122×73×43mm

イギリスのヴィクトリア期の茶箱。中が2つに分れていて、それぞれ小さな象牙のつまみのついた蓋の下に錫箔を張った茶入れがある。この手のものはこのごろ流行りの西洋骨董屋さんに良く並んでいるけれど、そういったたぐいにくらべてこれは比例もいいし、それになかなか品が良いとおもうのは、思い入れのゆえか。鍵がかかるようになっている。
220×130×130mm

11.
Two wood-carved boxes from France. The smaller one with roses is for stamps. It seems to be a touristic souvenir with the place and name printed.
The larger one is designed as a trunk. It is laboriously carved. This might be used as a cigarette case.
In collecting many boxes for a long time, there are occasions that similar boxes like these meet their relatives.

ぴっちりして隙間が無いということもその印象を強めよう。火に関してはまったくの言い伝えに過ぎず、いろいろな方がこれについて調べて書かれていることをみても、ほとんど関係ないようだ。着火温度、発火温度が異常に高いということはなく、たしかに伝説のようなものらしい。金庫の内箱などに桐が使ってある事などがこの印象を助けるのだろうが、あれはやはり狂いが少なくきっちりしているのと、先述の吸湿性の問題に若干かかわってのことであろう。それにしてもこういったことから、桐箱には貴重品の容れものというイメージが定着したようである。(古い物では貞享2年(1685)、加賀藩主前田綱紀公が寄進した有名な東寺百合文書をおさめる箱が桐箱で、12㎜厚の桐板を用いている。)

もっとも、桐の春日部も桐材そのものが自前で生産出来なくなっている訳で、そこでむしろ今後は伝統の技術を生かして他の材料も使っていろいろな箱を造っていこうとしているらしい。その試みのひとつが今回のデザイナーたちによる一連の作品という訳だが、これもまた影山さんの話によれば、その切替えが大変だったらしい。何しろ桐はソフトな材料だから、工作機械がすべて軟材用のもので、そこへ前回、今回と同様な試みをしたときに硬材を持ち込んだので、皆、機械を駄目にしてしまったという笑えないエピソードもあったようだ。で、今回の試みは2回目ということになる訳だが、ホット・スタンプにシルクスクリーン印刷、生地を見せるウレタン塗装、生地をつぶしてのラッカー塗装と、仕上げも各種試みられており、また形態としても従来の箱のイメージに

11.

フランスの木彫りの箱二題。小さいばらの文様のは切手入れになっていて、お土産品らしく地名が入っている。大きいほうはトランクふうの仕立てだが、それにしても良く丹念に彫り込んだものである。こちらはシガレット入れだろう。こうやって箱を蒐めているとこんなふうに何となく互いに親戚めいたものが出合うことがよくある。
大124×73×23㎜
小67×46×20㎜

12.
These kinds of boxes with copy-picture type prints used to abound in my childhood. Both boxes are composed of cedar planks, printed and then varnished. When I found them at a morning market, they were so untidy, and so I got them at a very cheap price. I washed them and took off the varnish very carefully, and they turned out to be very colorful. I use the larger one for keeping postcards.

13.
This is a varnished craft of paulownia. A flower of peony is painted with the sweep of a brush. Today, boxes are made of plastics, But in those days, boxes must have been made of wood or of lacquered papier-mâché. Then, they came to be made of celluloid. But I feel attached to these one-time prevalent ones, and, from time to time, take out to touch them.
The inside is two-layered for use as a sewing kit.

15.
Nest boxes of cedar with inlays. Some of the boxes made by inlaying various types of wood taking advantage of their texture and color are so elaborated. I often admire such tireless work. Bellflowers and eulalias are painted in various forms. This pattern is often seen on boxes of lacquered papier-mâché..
I found them carelessly displayed at a morning market and later polished them.

14.
The box with narcissuses is made of a camphor tree.
The other one with autumn plants is of paulownia.
Each piece of box of hand drawn prints has unresisting charm of its own as compared to the mass-produced goods of today.
Lyricism of the pictures on the wood box and of lacquered papier-mâché alike truly strikes a chord of Japanese sentiment.
I feel convinced by an art historian Toyomune Minamoto who expounded that the essence of Japanese aesthetic beauty lies in autumn plants.

16.
A box with raised pictures that might have been prevalent in the Taisho Era. (1912 — 1926).
The inside pasted with girlish scarlet red paper is also charming.
The casual encounter with this box at a street corner might have put fire on my extraordinary zeal for collecting boxes. Until then, I had some boxes on hand which were kept with no particular intention. But after the moment I bought this box, I came to feel as if I'm obligated to collect boxes.

17.
The master of a shop in Kyoto who always pleases me with delicious dishes makes sake cups himself, very tempting ones. When I go to his place, I often ask for a cup. He sometimes gives me one in a box of his choice. His drawing as well as writing are so fine that I feel excited at the first sight of the exterior, before seeing what is inside.
The model of this drawing, of course, is that of myself.

18.
The package used by a sweet shop in Nagoya sometime ago. It is a box made of paulownia. It bulges out in the shape of a drum. The handle is made of bamboo.
I don't usually collect packages but in case they are used as independent boxes like this, I add them to my collection.
This was given to me by the person who had treasured it.
Everyone has one or more boxes. Asking the transfer of such boxes is one technique of box collection.

捉われない自由な形のものが求められている。以前にも北海道立美術館が同様な企画で「あそびの木箱」展というのをやって、そのカタログに序文を求められたことがあったが、私は箱というのはデザイナーにとってきわめて魅力的なテーマではないかと思っている。というのも、他のものと違って、これはその形を造形するとともに、そこに小さいとはいえひとつの内部空間を創り出す仕事だからである。その空間の意識が薄いと、やはり「箱」としての魅力は半減してしまうというか、箱が箱でなくなってしまうのであって、形態や仕上げに凝らされる工夫がそうした空間の創成にどんなふうにかかわるかというところが箱のデザインのポイントであろう。デザインとしてはいろいろ努力していても、何だか箱としての魅力がさっぱり無いということがよくある。私のような箱気ちがいにとってはその箱としての魅力ということだけが問題なのだが、しかしそれはあながち箱にとりつかれた人間の妄執のなせる業ではなくて、やはりそれが箱というものの本質なのではないだろうか。この点、とくにデザイナーが手掛けたというものでもない一般に生産されている大量生産の箱のなかにも、箱としての魅力をたたえた存在を発見することが往々にしてある。たぶん箱という主題は面白いことは面白いが、またデザインとしては一筋縄ではいかない、なかなか難しいジャンルなのである。

こういう仕事はしかしひとりデザイナーがいても駄目であって、それに加えるに良い腕をもった技術者がいる必要がある。いまは全体として大量生産化とともに簡単にやれる工程が多用され、その結果として出来上るもの

12.

昔はこんなふうに一種のうつし絵ふうのプリントをした箱がよくあったものだ。どちらも杉板を組んで出来ていて、プリントをしたうえにニスが掛けてあったものである。朝市で求めたときはどちらも汚なくてそれこそただ同然で貰って来たのだが、ニスの部分を丹念に洗い落し剝ぎ落していったら実に味のある箱になって来た。大きい方は絵葉書入れに使っている。
大207×145×80mm
小117×84×47mm

13.

これはニス塗りの桐細工の箱。牡丹が一輪さっと描き上げてある。いまはプラスチック全盛になってしまったけれど、昔は木でなければ一閑張というのが箱の常道だった。それがセルロイドになりプラスチックになったのである。何だかこうしたほんの少しまえの箱たちがいとおしい気がして、ついついこの種の箱を手にしてしまう。中は小型の裁縫箱の造りで、懸子がついている。
210×147×85mm

14.

水仙の絵のほうは楠の箱。秋草の絵の方は桐で出来ている。こうやってひとつひとつ絵が描かれた箱は、いま出来の大量生産のものにくらべて、それぞれにひとの心をそそってやまないものがある。それに一閑張の箱もそうだが、こうした箱に描かれた絵の抒情は、まさに日本人の心の琴線に触れるものなのである。美術史家源豊宗が日本の美の本質は秋草にありと喝破したのが思い出される。
大145×115×50mm
小100×66×35mm

17.

京都でいつも美味しいものを喰べさせていただくお店の御主人は、いかにも酒を注いだらうまそうだというぐい呑みを焼かれる。で、お店に伺うとこれ頂戴と言ってポケットに入れて帰るのがしょっちゅうなのだが、ときにこのように作品を択んで箱に入れて下さる。この方は絵も字も素敵で、中味もいいが外の箱書でまず嬉しくなってしまう。人物像のモデルはもちろん私である。
大85×85×90㎜
小70×70×85㎜

18.

名古屋のお菓子屋さんが昔使っていたパッケージ。桐の箱だが、胴がふくらんで彭形の良い形になっている。つまみは竹。私はいわゆるパッケージは原則として蒐めないし、またその気にならないのだが、こんなふうに箱として自立しているときは、もちろんコレクションに加える。これはある人が大事に物入れにしていたのを譲り受けたもの。誰でも何かひとつは、そうした箱を持っているもので、それをうまく貰い受けるのが、また蒐集のこつである。
100×100×83㎜

の質が低下するという現象が一般にみられる。桐箱の表面の仕上げにしても、仕上げカンナでやったものが、いまはサンドペーパーを用いるといった具合である。そういった全般的な傾向のなかで、春日部でも若いひとたちのなかに伝統的な技法を守って本来の質を保っていこうとしているひとが何人かは出て来ているという。たしかに桐の生地の美しい味合いは、サンドペーパー仕上げで出て来ない。そうした人たちの努力と、すぐれたデザイナーの創作とがうまく組合わされることで、伝統の「箱」に新しい生命が吹き込まれたらすばらしいとおもうのである。箱というのは不思議なもので（もっともあらゆるクラフトがそうと言えばそうなのだが）、隅の二つの面の直角に交わるところのおさまりかた、あるいは蓋と身の重なり具合といった微妙な部分がかえってデザイン全体の生命の鍵となっているようなところがある。それをおさえるのは、やはり箱をつくるというプロセスを知り抜いていなくては叶わないところで、そういった視点から見ると現代の「箱」にはまだまだ工夫の余地がたくさんあるように思われる。今後こうした試みが幾度も行なわれるなかで、デザイナーと工人の理想的な協同、あるいは新しいタイプのデザイナーが誕生していくことに期待をかけたいものである。

16.

15.

杉の象嵌の箱。容れ子になっている。いろいろな木の色や質感を利用して文様をつくるこの手の象嵌の箱には実に手の込んだものがあって、良くこんなに手間をかけてと感嘆させられることが多い。これはごく庶民的なものだけれど、同じ桔梗とすすきを描きながら大小それぞれに柄を変えている。一閑張の手箱に良く見る図柄である。これも朝市に哀れな姿で転っていたのを直してきれいにしてやったもの。
大113×86×40㎜
小95×68×33㎜

押絵の箱。大正あたりというところか。いかにも女の子らしい紅色の紙を貼った内側の状態もいい。この箱に偶然、街角で出会ったことから、ぼくの異常な箱狂いがはじまったと言っていいようだ。それまでも箱は何となく手もとに溜っていたのだが、これを手に入れたとたん、箱を蒐めなくてはならぬという義務感のようなものに取りつかれてしまったのである。
160×130×60㎜

Wood that Makes Boxes, Boxes Made of Wood

What was the first box in Japan made of? Certainly it did not seem to have been made of pottery or metal. Mr. Akio Haino of the National Museum of Kyoto once said in his article on the history of lacquer ware in Japan that the history of box-making in Japan could have begun with leather. Although he admits, there is no definite proof, the majority of containers of the Nara period (710—794) preserved today are made of leather.
Considering the engineering work involved in making boxes made of wood or bamboo, Mr. Haino assumes that the emergence of wooden an boxes may be a little later than the leather packing. In contrast, leather containers could have had an earlier beginning since leather can be sewed. The reason for wooden in boxes to later become the mainstream of packing could be attributed to its shape and its strength of protecting its contents. Mr. Haino gathers that the tendency to value decoration may be a reason to favor wood boxes.
There seems to be two types of boxes: one to be always exposed before viewer's eyes, and the other to be kept in a closet or warehouse simply as a container.
The former is elaborately made and usually decorated either by lacquering or carving. The latter, even if costly made, is plain with no showy decoration.
If the packing is seen as an outstanding entity, then, the one with better potential for decoration will be favored. Leather packages could be decorative as seen in the beautiful filigreed pieces of the Yi Dynasty of Korea and in other treasures of old time China.
However, leather or even metal and pottery cannot compare with wood as far as the degree of expressive freedom is concerned. I have collected quite a few kitsch sort of boxes, and am often amazed with the variety of expression these boxes have.
The wooden boxes described in this book are simple, traditional boxes used by the common people.
Those contemporarily designed boxes have diverse expressions. When was this kind of technology developed?
Among the divine treasures at Ise Shrine is a well known withe box. This box is made of withes tied with silk threads. The method appears to be the most primitive way of box making.
In the Nara period, joining planks by mortising, nailing and using a kind of adhesive to make a box seems to have been established. However, in those days, boxes were used only by high-ranking people.
The boxes, apparently, were introduced into the lives of common people in the following — Heian period (794—1192).
It was discovered from the scrolls of the Kamakura period (1192—1333) the common people used containers made of wood blocks or curved boards. Even so, each family had only a limited number of containers, for people in those days had very few things to store, and the making of boxes required a very complicated craftwork.
It was during the Edo period (1603—1868) that wooden boxes were mass-produced and was deeply-rooted in the life of common people. Behind this development was the technical innovation of woodworking during the preceding Muromachi (1392—1573) and Azuchi Momoyama (1573—1598) periods.
In the paintings and wood block prints of the Edo period, evidence is seen in various shapes of wood boxes made and used by the common people.
The cultural maturity of the Edo-period townspeople led to the sophisticated craftsmanship and masterly performances. This also affected box making.
Thus, box culture flourished among the high-rank people already in the Kamakura and Muromachi periods, but it was not until the Edo era that it flourished among the ordinary people.
A document kept in Kasukabe city to prove the townspeople in the art of making wooden boxes in the mid-Edo era is one such evidence. Located along the main road connecting Tokyo and Nikko, carpenters and cabinetmakers involved in building shrines and temples in Nikko seem to have gathered in the city.

Associate Professor, The University of Tokyo
Tadashi Yokoyama

I have thought that this city, famous for its production of paulownia wood chest of drawers, started and expanded and in the process started to make smaller items using cuttings of wood.
But the fact was reverse.
According to a researcher of Industrial Technology Institute of Saitama Prefecture, the location of the city served favorably for the development of the industry. Box-making industry in the area was at its prime during the Taisho period and the early years of Showa. The city's development as a box producing town mainly depended on the availability of paulownia trees, in fact, self-supplied paulownia wood was used until the early Showa era. When local supply was exhausted, paulownia was imported from all over Japan, and from overseas as well.
The paulownia is used because the tree is widely available in Japan, the wood is very light (the lightest of all kinds of wood in Japan), and it has a soft and beautiful texture when seen at first glance. But these are not all the factors to attract people to choose boxes made of paulownia.
I feel it is perhaps because of a legend that a box made of paulownia wood is fire-resistance and water-resistance.
The legend says that clothes put in a paulownia cabinet did not get wet in a flood, or anything kept in a paulownia chest escaped from burning in a fire. Indeed, paulownia wood has low hygroscopicity and water absorption.
Because of small contractibility, (similar to that of other wood with the same gravity) a paulownia box may give an impression that the lid and body are tighty closed. It is, therefore, reasonable that paulownia boxes are strong against water and humidity.
Nontheless, its strength against fire is entirely illusory.
There is no scientific ground to support this legend. A cashbox with paulownia backing may give such an impression. These properties of paulownia wood seem to have caused people to think that such containers made of the wood are for storing valuable articles.
As Kasukabe city can no longer depend on its self-supplied paulownia wood, box manufacturers began to make most of their traditional techniques in making boxes out of various other kinds of wood. The series of works by designers presented here are part of such efforts.
(The switchover of the material from paulownia to other harder materials was damaging to machines originally made for soft wood.)
The designers made various attempts in their experimental works: hot-stamping, silk screening, urethane coating showing the surface of wood, lacquering and swearing off the surface. Forms are also freely created.
The design of boxes must be challenging and exciting as it involves the creation of both form and interior space, though it is small. Whatever elaborate work is applied to the form and exterior of the box, a box cannot be called as such unless it has a well-considered space inside. I often come across well-designed boxes but lacking in charm. On the contrary, there are mass produced boxes which are charming enough as boxes. Perhaps, the box belongs to a difficult genre of design.
A charming box cannot possibly be made by designers alone. There should be competent craftsmen.
In Kasukabe, there are some young people who are positively preserving traditional methods against mass-production process. When their strenous work is to be combined with the creativity of excellent designers, fresh air would be blown into the traditional box.
A box is a wonder: the right-angled corner where two facets meet, or the delicate joint of the lid and body would illustrate the key of the total design. Both designers and craftsmen must master the box making process in order to bring forth a truly beautiful box into view.
From this standpoint, there is still much to be developed before the box of today is created. Through number of design trials, an ideal cooperation between designers and craftsmen will emerge, including new types of designs.

広橋桂子

KEIKO HIROHASHI

1.............Personal Box • Slanting Box
material used : Paulownia

KEIKO HIROHASHI
広橋桂子

1............パーソナル・ボックス・斜箱
使用素材............桐

1

2 Personal Box • Pocket
material used : Paulownia
surface finish : Clear Lacquer

3 Personal Box • Building Blocks
material used : Paulownia
surface finish : Clear Lacquer

KEIKO HIROHASHI
広橋桂子

2 パーソナル・ボックス・ポケット
使用素材……桐
表面処理……クリアラッカー塗装

3 パーソナル・ボックス・積み木
使用素材……桐
表面処理……クリアラッカー塗装

2

3

4............Gift Box

material used : Paulownia
surface finish : Clear Lacquer

KEIKO HIROHASHI

広橋桂子

4............ギフト・ボックス

使用素材..........桐
表面処理..........クリアラッカー塗装

4

fine time!

5……Gift Box
material used : Paulownia
surface finish : Clear Lacquer

6……Personal Box • Magic
material used : Paulownia

KEIKO HIROHASHI
広橋桂子

5……ギフト・ボックス
使用素材……桐
表面処理……クリアラッカー塗装

6……パーソナル・ボックス・マジック
使用素材……桐

5

6

7........Gift Box • Knockdown
materials used : Paulownia & Cloth Belt
surface finish : Clear Lacquer

8-9........Personal Box • Table and Stool
materials used : Paulownia & Stainless Steel

KEIKO HIROHASHI
広橋桂子

7........ギフト・ボックス・ノックダウン
使用素材........桐・布ベルト
表面処理........クリアラッカー塗装

8-9........パーソナル・ボックス・テーブル&ストゥール
使用素材........桐・ステンレス

7

8

9

10.............Personal Box • Book
material used : Paulownia
surface finish : Clear Lacquer

11.............Personal Box • Wingbeat
material used : Paulownia

KEIKO HIROHASHI
広橋桂子

10……パーソナル・ボックス・ブック
使用素材……桐
表面処理……クリアラッカー塗装

11……パーソナル・ボックス・はばたき
使用素材……桐

10

12……Personal Box • Flow
material used : Katsura
surface finish : Color Urethane

13……Personal Box • Beans
materials used : Katsura & Stainless Steel
surface finish : Color Urethane

14……Personal Box • Multiplation
material used : Katsura
surface finish : Color Urethane

KEIKO HIROHASHI

広橋桂子

12……パーソナル・ボックス・流れ
使用素材……桂
表面処理……カラーウレタン塗装

13……パーソナル・ボックス・ビーンズ
使用素材……桂・ステンレス
表面処理……カラーウレタン塗装

14……パーソナル・ボックス・増殖
使用素材……桂
表面処理……カラーウレタン塗装

12

13

14

15.............Personal Box • Breezing Sounds
material used : Katsura
surface finish : Color Urethane

16.............Personal Box • Swing
material used : Katsura
surface finish : Color Urethane

17.............Personal Box • Zig Zag
material used : Katsura
surface finish : Color Urethane

18.............Personal Box • Step
material used : Magnolia Hypoleuca
surface finish : Color Urethane & Hand-drawing Pattern

KEIKO HIROHASHI

広橋桂子

15……パーソナル・ボックス・風声
使用素材……桂
表面処理……カラーウレタン塗装

16……パーソナル・ボックス・スウィング
使用素材……桂
表面処理……カラーウレタン塗装

17……パーソナル・ボックス・ジグ・ザグ
使用素材……桂
表面処理……カラーウレタン塗装

18……パーソナル・ボックス・ステップ
使用素材……朴
表面処理……カラーウレタン塗装・手描きパターン

15

17

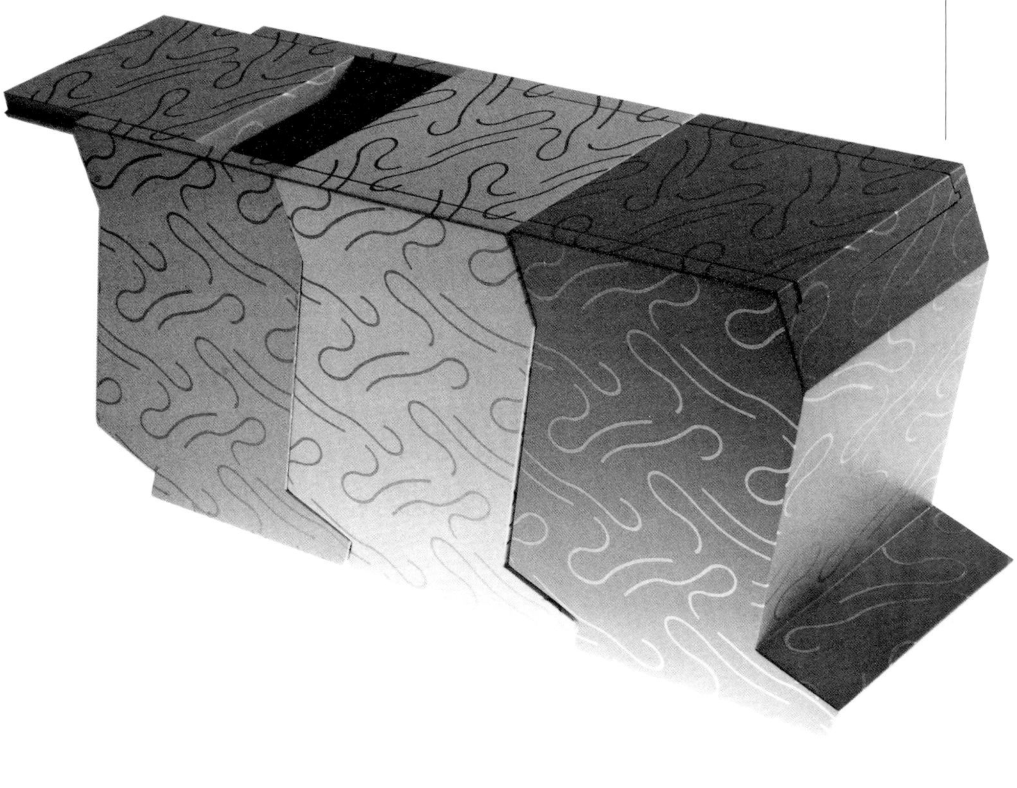

18

16

杉村敏男

TOSHIO SUGIMURA

1.........Performance Box
material used : Plywood
surface finish : Lacquer

TOSHIO SUGIMURA
杉村敏男

1.........パフォーマンス・ボックス
使用素材.........合板
表面処理.........ラッカー塗装

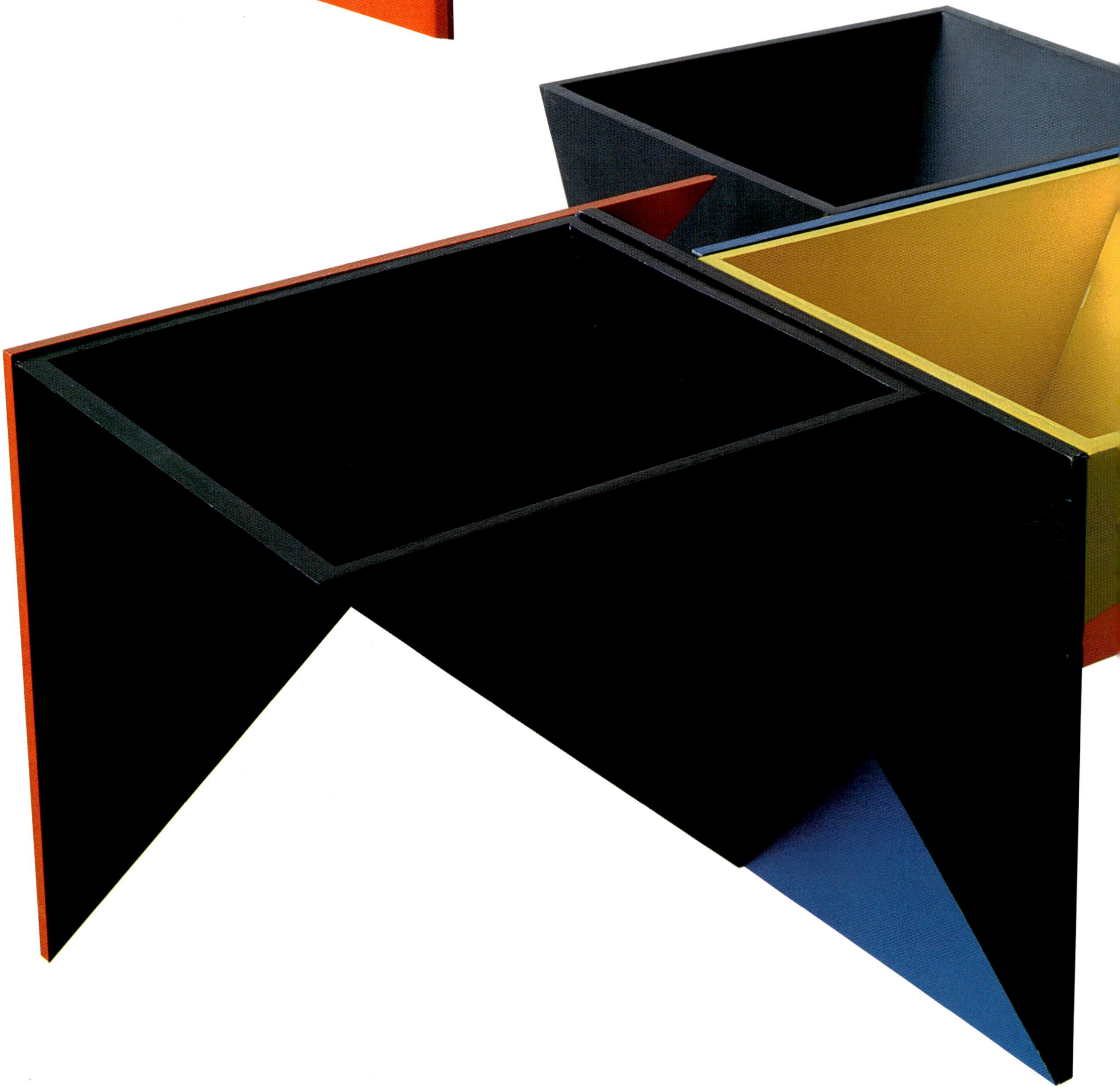

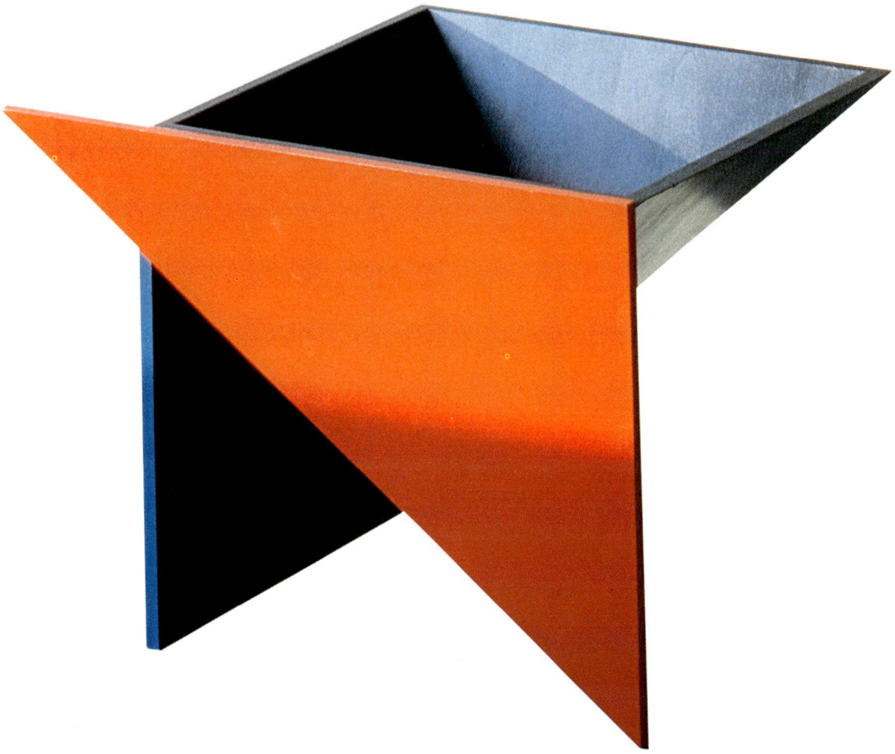

1

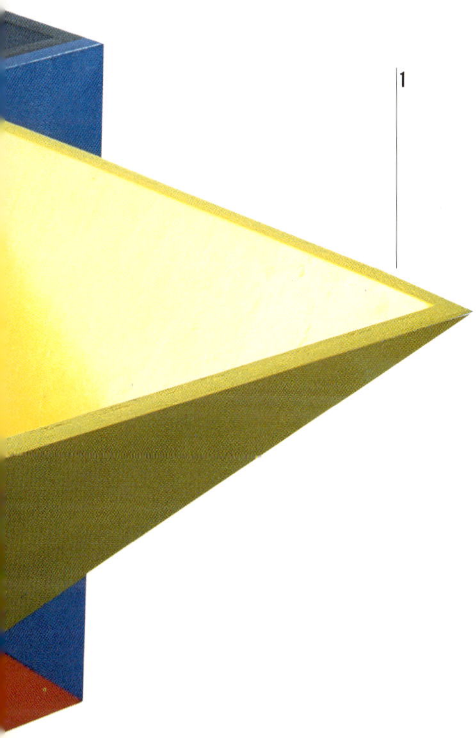

TOSHIO SUGIMURA
杉村敏男

2……キュービック・ボックス2
使用素材……チーク・ステンレス

3……キュービック・ボックス3
使用素材……チーク・ステンレス

4……ベジタブル・ボックス
使用素材……桐
表面処理……シルクスクリーン印刷

5……ベジタブル・ボックス
使用素材……桐・杉
表面処理……シルクスクリーン印刷

2

3

VEGETABLE BOX

4

VEGETABLE BOX
Wood's Original Package Design Timely Box

5

6……Performance Box
material used : Teak Wood
surface finish : Lacquer

TOSHIO SUGIMURA
杉村敏男

6……パフォーマンス・ボックス
使用素材……チーク
表面処理……ラッカー塗装

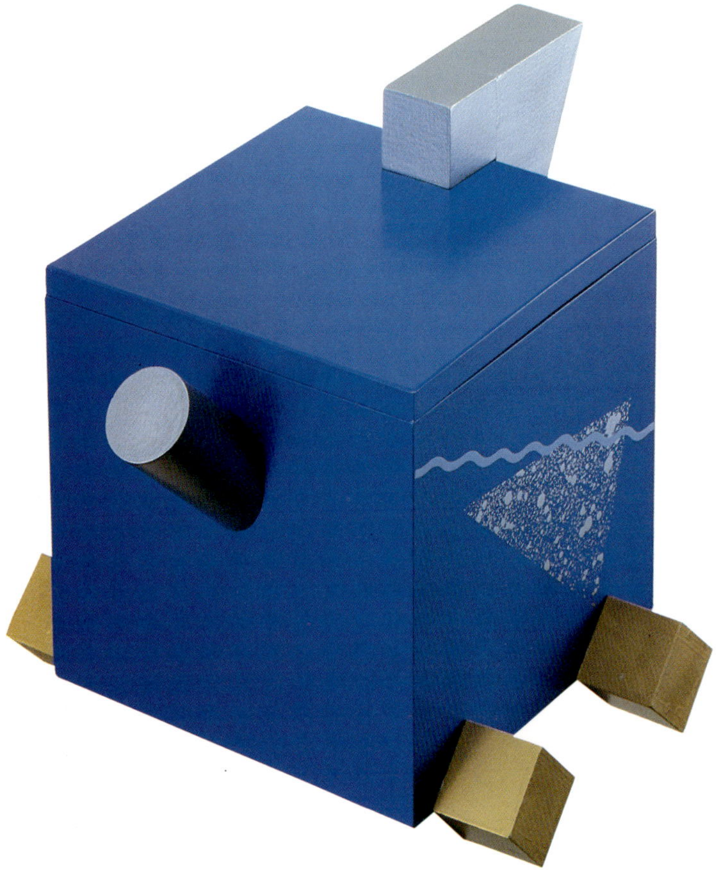

6

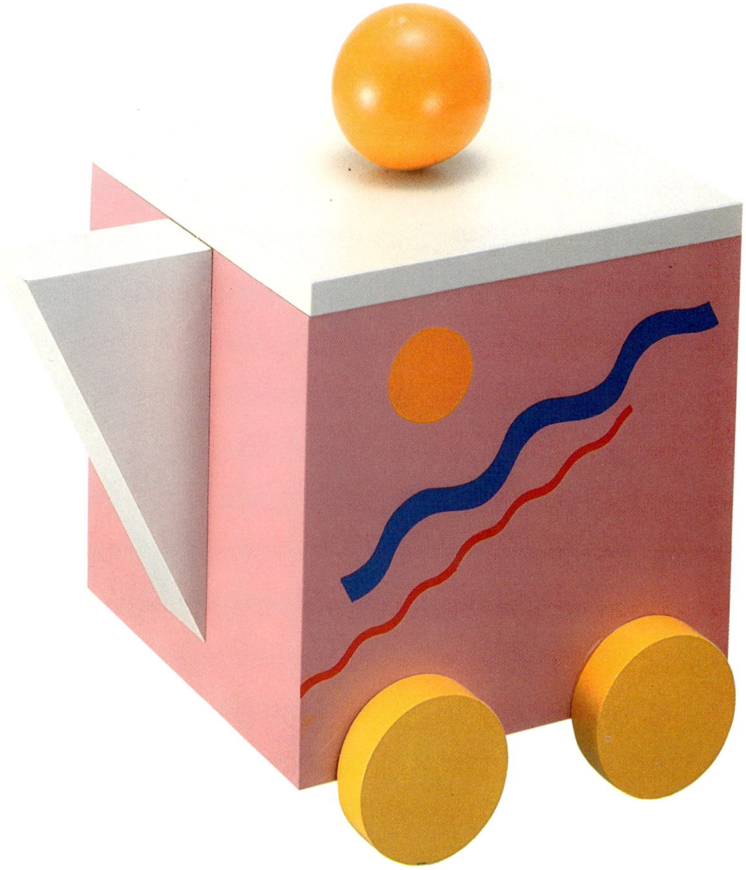

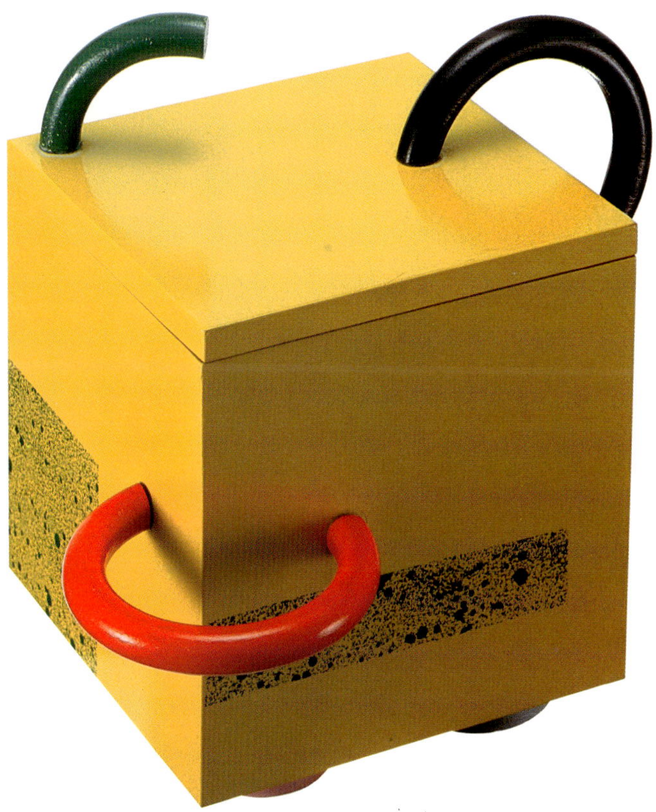

TOSHIO SUGIMURA
杉村敏男

7 行燈ボックス
使用素材……桐
表面処理……オイルステイン仕上げ

8 パフォーマンス・ボックス
使用素材……桐
表面処理……オイルステイン仕上げ

9 ジグザグ・ボックス
使用素材……桐

10 整理箱
使用素材……桐
表面処理……オイルステイン仕上げ

7

8

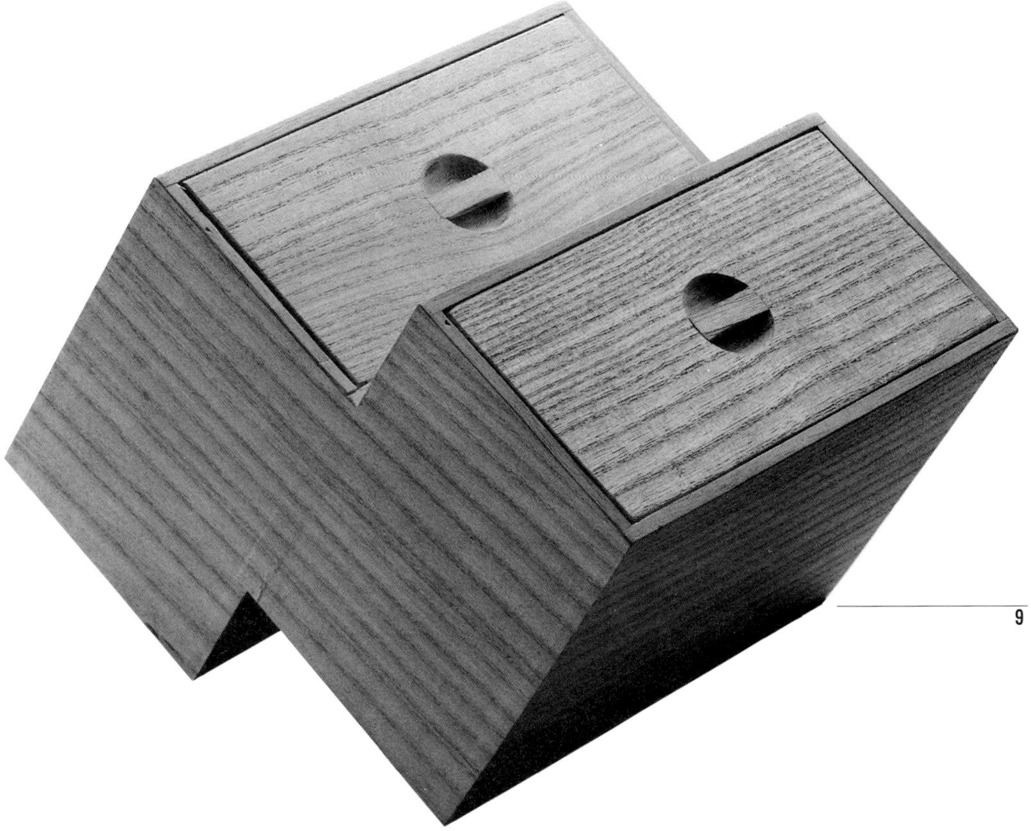

9

10

11.............Mistery Box
materials used : Japanese Oak & Paulownia

12.............Cubic Box 1
materials used : Paulownia & Teak Wood

TOSHIO SUGIMURA

杉村敏男

11………ミステリー・ボックス
使用素材………楢・桐

12………キュービック・ボックス1
使用素材………桐・チーク

11

12

13……Clock Box
materials used : Katsura & Japanese Cedar
surface finish : Oil Stain & Vanish

14……Joint Box
material used : Japanese Oak
surface finish : Oil Stain & Lacquer

15……Vegetable-shaped Casket
material used : Paulownia
surface finish : Silk Screen Process

16……Wine Gift Box
material used : Plywood
surface finish : Silk Screen Process

TOSHIO SUGIMURA
杉村敏男

13……クロック・ボックス
使用素材……桂・杉
表面処理……オイルステイン・ニス塗装

14……ジョイント・ボックス
使用素材……楢
表面処理……オイルステイン・ラッカー塗装

15……野菜の箱
使用素材……桐
表面処理……シルクスクリーン印刷

16……ギフト用ワイン・ボックス
使用素材……合板
表面処理……シルクスクリーン印刷

13

14

15

16

TOSHIO SUGIMURA
杉村敏男

17……**Fish-shaped Box**
material used : Paulownia
surface finish : Silk Screen Process

18……**Weekly Box**
material used : Paulownia
surface finish : Stain & Lacquer

19……**Joint Audio Box**
materials used : Teak Wood & Audio Kit
surface finish : Lacquer

17……**魚の箱**
使用素材……桐
表面処理……シルクスクリーン印刷

18……**ウィークリー・ボックス**
使用素材……桐
表面処理……ステイン・ラッカー塗装

19……**ジョイント・オーディオ・ボックス**
使用素材……チーク、オーディオ機具
表面処理……ラッカー塗装

17

18

U.G.サトー

U.G. SATO

1……Eroded Box・Horse
material used : Kalopanax
surface finish : Lacquer

U.G.SATO
U.G.サトー

1……蝕まれた箱・馬
使用素材……センノキ
表面処理……ラッカー塗装

1

2.........Kissing Box
material used : Linden Plywood
surface finish : Urethane

3-6.........Box without a Corner•Bat•Grasping Hand•Heart•Man & Woman
material used : Kalopanax
surface finish : Urethane

2.........キッシング・ボックス
使用素材.........シナ合板
表面処理.........ウレタン塗装

3-6.........角をとる箱・コウモリ・掴む・ハート・男と女
使用素材.........センノキ
表面処理.........ウレタン塗装

2

3

4

2

U.G.SATO
U.G.サトー

2………キッシング・ボックス
使用素材………シナ合板
表面処理………ウレタン塗装

3-6………角をとる箱・コウモリ・掴む・ハート・男と女
使用素材………センノキ
表面処理………ウレタン塗装

2

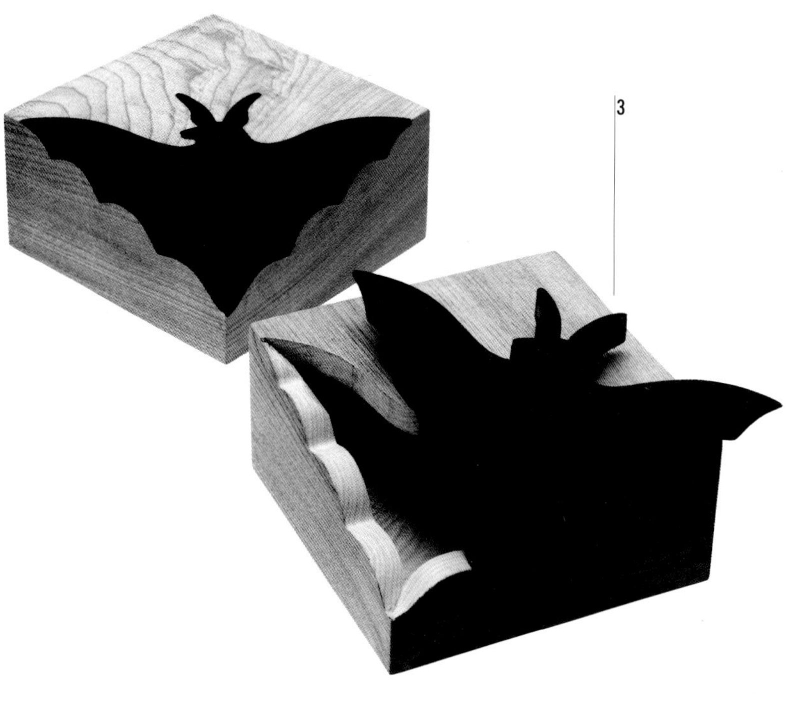

3

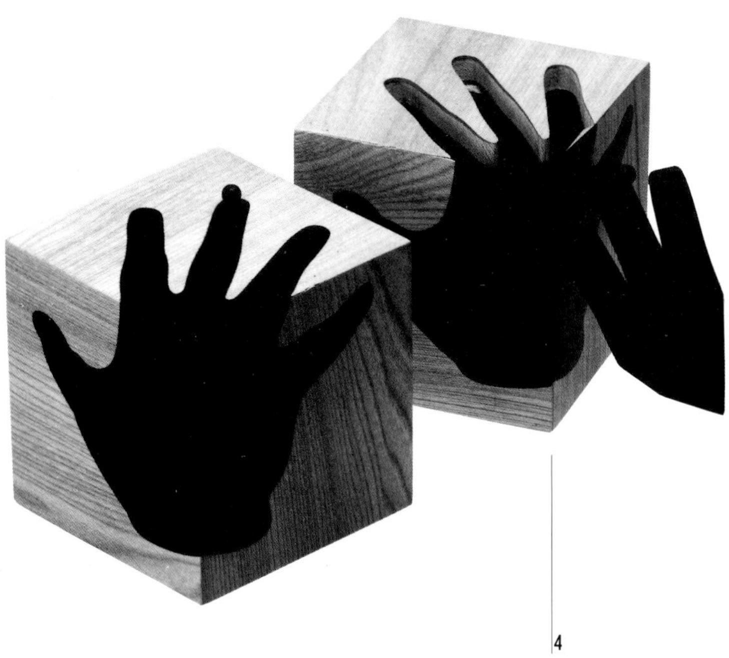

4

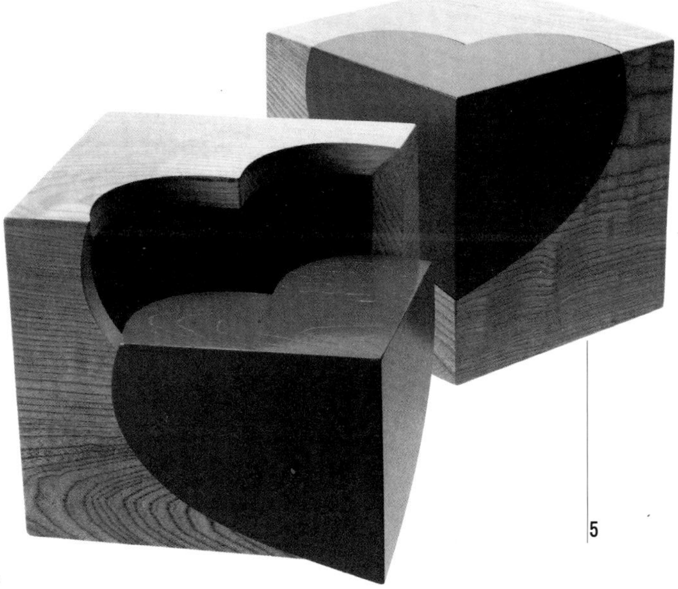

5

6

7 Walking Box
material used : Kalopanax
surface finish : Lacquer & Inlay

8 Relaxing Box
material used : Beech Wood
surface finish : Urethane (Partial)

9 Noisy Box
material used : Kalopanax
surface finish : Urethane

U.G.SATO

U.G.サトー

7 ウォーキング・ボックス
使用素材……センノキ
表面処理……ラッカー塗装・象眼

8 くつろぎの箱
使用素材……ブナ
表面処理……ウレタン塗装

9 さわがしい箱
使用素材……センノキ
表面処理……ウレタン塗装

7

8

10.............Stair Box
material used : Kalopanax
surface finish : Lacquer & Inlay

11.............Stair Box
material used : Linden Plywood
surface finish : Silk Screen Process

12.............Seesaw Box
material used : Linden Plywood
surface finish : Lacquer

13.............Swing Box
material used : Linden Plywood
surface finish : Silk Screen Process

U.G.SATO

U.G.サトー

10.........のぼりおりの箱
使用素材.........センノキ
表面処理.........ラッカー塗装・象眼

11.........のぼりおりの箱
使用素材.........シナ合板
表面処理.........シルクスクリーン印刷

12.........シーソー・ボックス
使用素材.........シナ合板
表面処理.........ラッカー塗装

13.........スウィング・ボックス
使用素材.........シナ合板
表面処理.........シルクスクリーン印刷

10

11

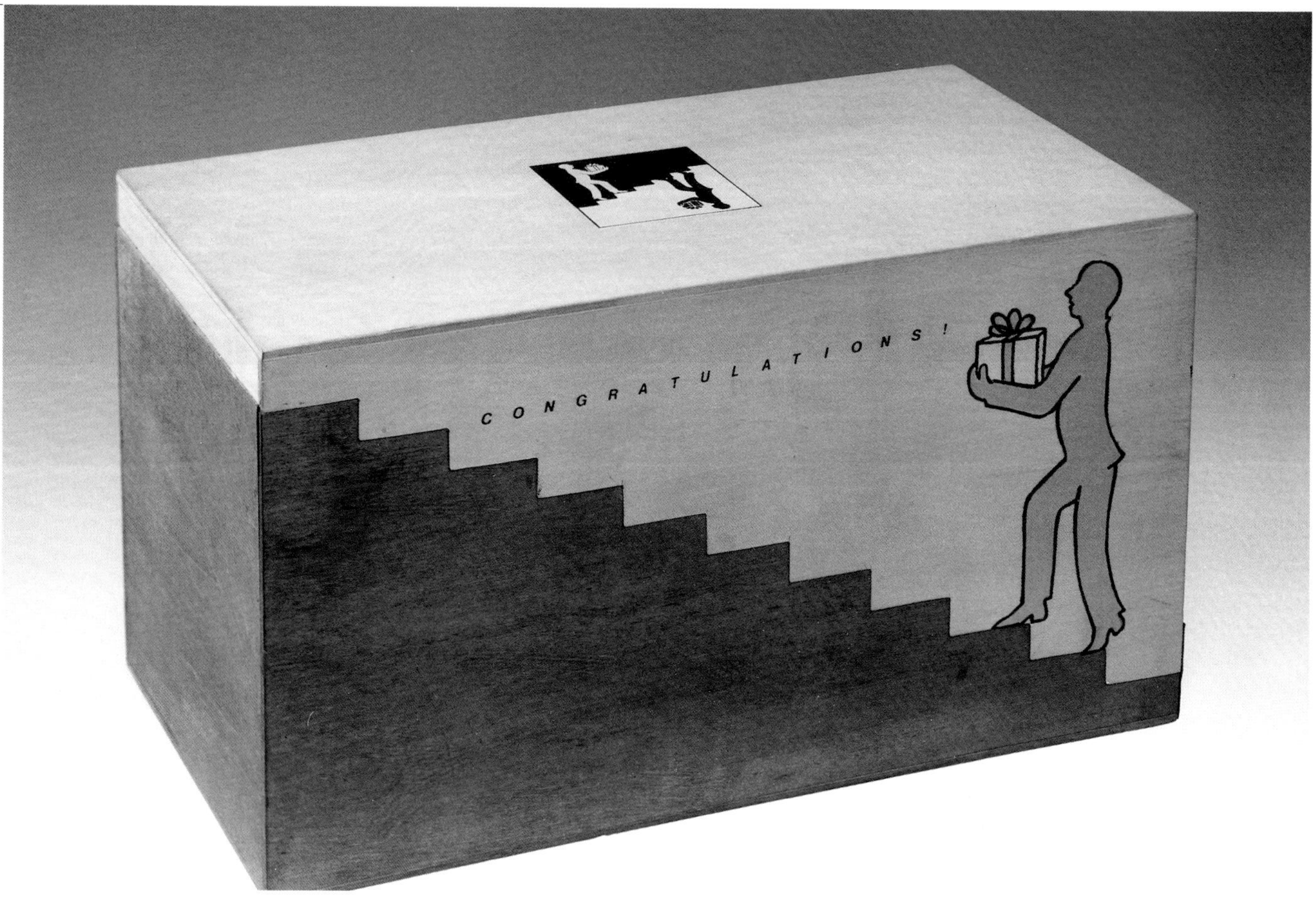

12

13

14……Flying Box
materials used : Kalopanax & Cactus
surface finish : Inlay

15……Frisky Box
material used : Linden Plywood
surface finish : Lacquer

16……Flying Box
material used : Linden Plywood
surface finish : Urethane (Partial)

U.G.SATO
U.G.サトー

14……飛び立つ箱
使用素材……センノキ・カクタス
表面処理……象眼

15……たわむれる箱
使用素材……シナ合板
表面処理……ラッカー塗装

16……フライング・ボックス
使用素材……シナ合板
表面処理……ウレタン塗装

14

15

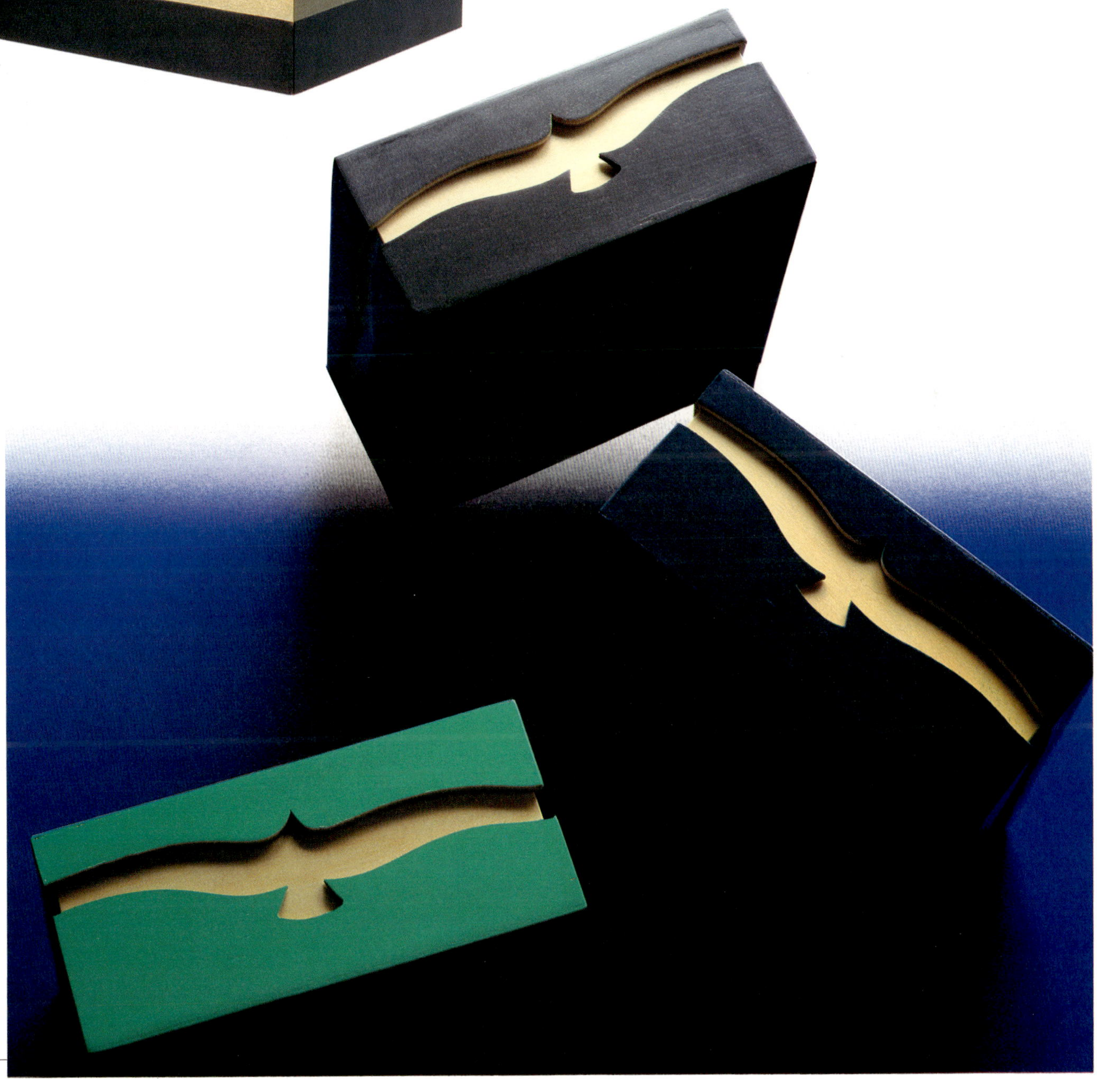

16

17……バード・ボックス
使用素材……シナ合板

18……持ち上った箱
使用素材……シナ合板
表面処理……ウレタン塗装

19……蝕まれた箱・オートバイ
使用素材……センノキ
表面処理……ラッカー塗装

20……蝕まれた箱・日本
使用素材……シナ合板
表面処理……ラッカー塗装

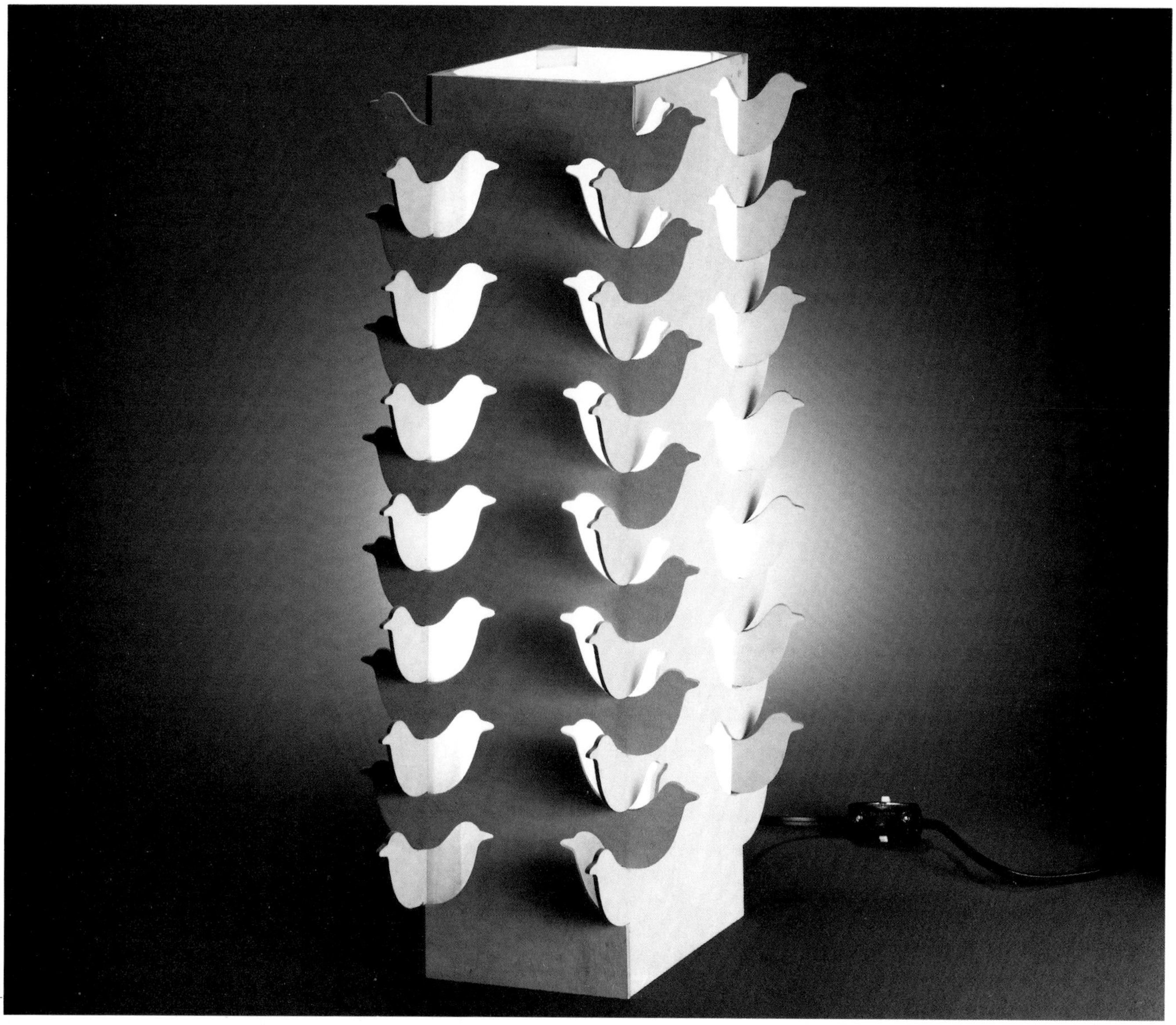

17

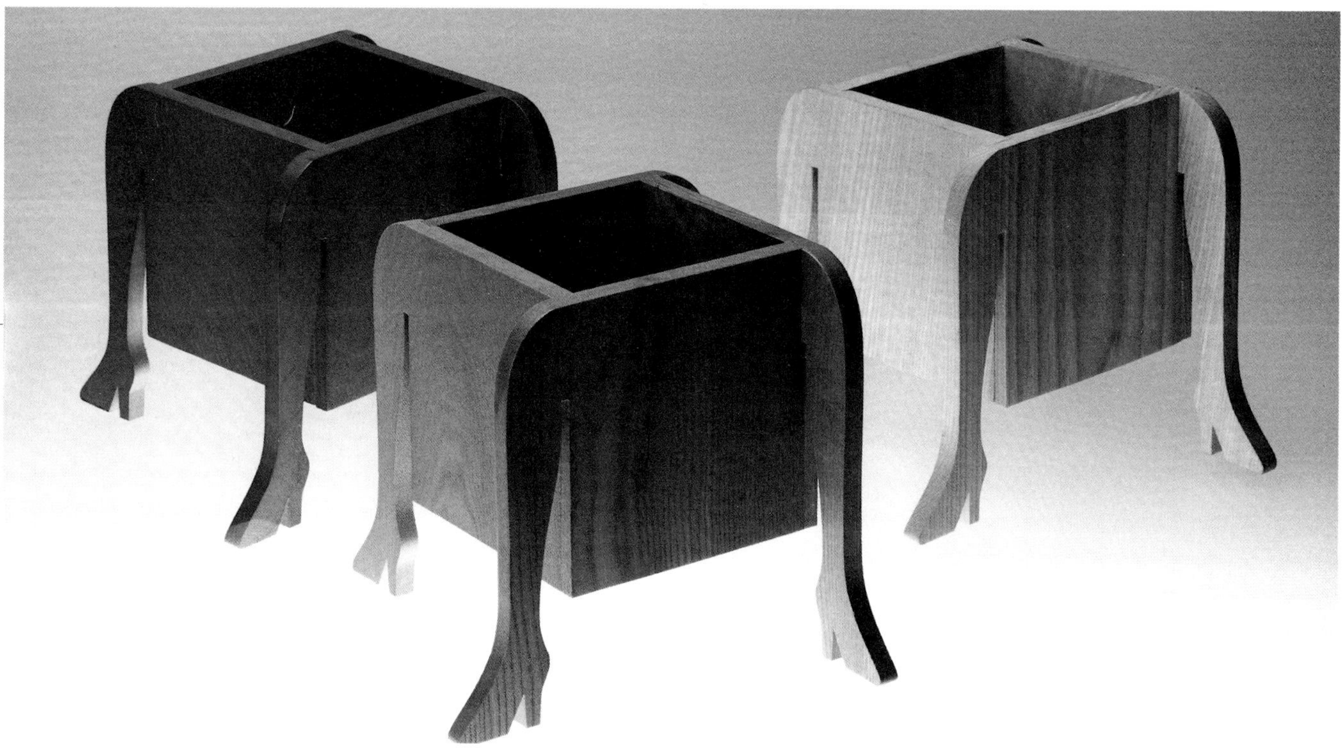

18

19

20

21 Eroded Box • World
material used : Linden Plywood
surface finish : Lacquer

U.G.SATO
U.G.サトー

21 蝕まれた箱・世界
使用素材 シナ合板
表面処理 ラッカー塗装

21

岡田宏三

KOZO OKADA

KOZO OKADA

岡田宏三

1……フレキシブル・ギフト・ボックス

使用素材……桧・杉・栂・バルサ・シナ合板・ゴム・小松石・和紙

表面処理……スライスウッド(左)・カットウッド(右)

1……Flexible Gift Box

materials used : Japanese Cypress, Japanese Cedar, Japanese Hemlock, Balsa, Linden Plywood, Rubber, Stone & Japanese Paper

surface finish : Slice Wood (Left), Cut Wood (Right)

1

2 Flexible Gift Box
materials used : Balsa, Linden Plywood & Rubber
surface finish : Cut Wood

3 Jam Bottle Case
materials used : Japanese Cypress, Japanese Oak, Linden Plywood, Balsa & Cotton Cord
surface finish : Slice Wood (Up), Cut Wood (Down)

KOZO OKADA
岡田宏三

2 フレキシブル・ギフト・ボックス
使用素材……バルサ・シナ合板・ゴム
表面処理……カットウッド

3 ジャム・セット
使用素材……桧・楢・シナ合板・バルサ・綿ひも
表面処理……スライスウッド(上)カットウッド(下)

2

JAM

KOZO OKADA
岡田宏三

4……Stationery Case
materials used : Japanese Cypress, Cedar, Japanese Hemlock, Japanese Oak & Rubber
surface finish : Slice Wood

5……Stationery Case
materials used : Japanese Hemlock, Balsa & Rubber
surface finish : Cut Wood

4……ステイショナリー・ケース
使用素材……桧・杉・栂・楢・ゴム
表面処理……スライスウッド

5……ステイショナリー・ケース
使用素材……栂・バルサ・ゴム
表面処理……カットウッド

4

6……Apparel Holder
materials used : Balsa, Linden Plywood, Rubber & Acetate Film
surface finish : Cut Wood

7……Flexible Gift Box
materials used : Balsa, Linden Plywood, Japanese Cypress, Japanese Cedar, Japanese Hemlock & Rubber
surface finish : Cut Wood, Slice Wood

8……Flexible Gift Box
materials used : Japanese Cypress, Japanese Cedar, Linden Plywood & Rubber
surface finish : Slice Wood

KOZO OKADA
岡田宏三

6……アパレル・ホルダー
使用素材……バルサ・シナ合板・ゴム・アセテートフィルム
表面処理……カットウッド

7……フレキシブル・ギフト・ボックス
使用素材……バルサ・シナ合板・桧・杉・栂・ゴム
表面処理……カットウッド・スライスウッド

8……フレキシブル・ギフト・ボックス
使用素材……桧・杉・シナ合板・ゴム・真鍮リング
表面処理……スライスウッド

9……フレキシブル・ギフト・ボックス
使用素材……ヒバ・シナ合板・真鍮リング・ゴム
表面処理……スライスウッド

6

7

9.........Flexible Gift Box

materials used : White Cedar, Linden Plywood, Brass Ring & Rubber
surface finish : Slice Wood

8

9

10............Dreamy Pocket
materials used : Linden Plywood, Tamo, Cotton-linen & Cotton Cord
surface finish : Stripe Wood & Lacquer

11............Roll Box
materials used : Paulownia Plywood, Paulownia, Cotton-linen & Cotton Cord
surface finish : Stripe Wood & Silk Screen Process

KOZO OKADA
岡田宏三

10……ドリーミー・ポシェット
使用素材……シナ合板・タモ・綿麻布・綿ひも・ゴム
表面処理……ストライプウッド・ラッカー塗装

11……ロール・ボックス
使用素材……桐合板・桐・綿麻布・綿ひも
表面処理……ストライプウッド・シルクスクリーン印刷

10

12......ステイショナリー・ボックス
使用素材.........桐合板・桐・綿麻布
表面処理.........ストライプウッド

12

筆
FUDE

13........Cross Box
materials used : Paulownia Plywood, Paulownia & Japanese Paper
surface finish : Stripe Wood

14........Apparel Holder
materials used : Paulownia Plywood & Paulownia
surface finish : Stripe Wood & Silk Screen Process

15........Apparel Holder
materials used : Balsa & Brass Ring
surface finish : Cut Wood

KOZO OKADA
岡田宏三

13……クロス・ボックス
使用素材……桐合板・桐・和紙
表面処理……ストライプウッド

14……アパレル・ホルダー
使用素材……桐合板・桐
表面処理……ストライプウッド・シルクスクリーン印刷

15……アパレル・ホルダー
使用素材……バルサ・真鍮リング
表面処理……カットウッド

13

MEN'S
shirts

16.........Dreamy Pocket
materials used : Linden Plywood, Paulownia & Cotton-linen
surface finish : Parallel Wood & Silk Screen Process

17.........Parallel Box
materials used : Linden Plywood, Paulownia & Cotton-linen
surface finish : Parallel Wood & Silk Screen Process

KOZO OKADA
岡田宏三

16.........ドリーミー・ポケット
使用素材.........シナ合板・桐・綿麻布
表面処理.........パラレルウッド・シルクスクリーン印刷

17.........パラレル・ボックス
使用素材.........シナ合板・桐・綿麻布
表面処理.........パラレルウッド・シルクスクリーン印刷

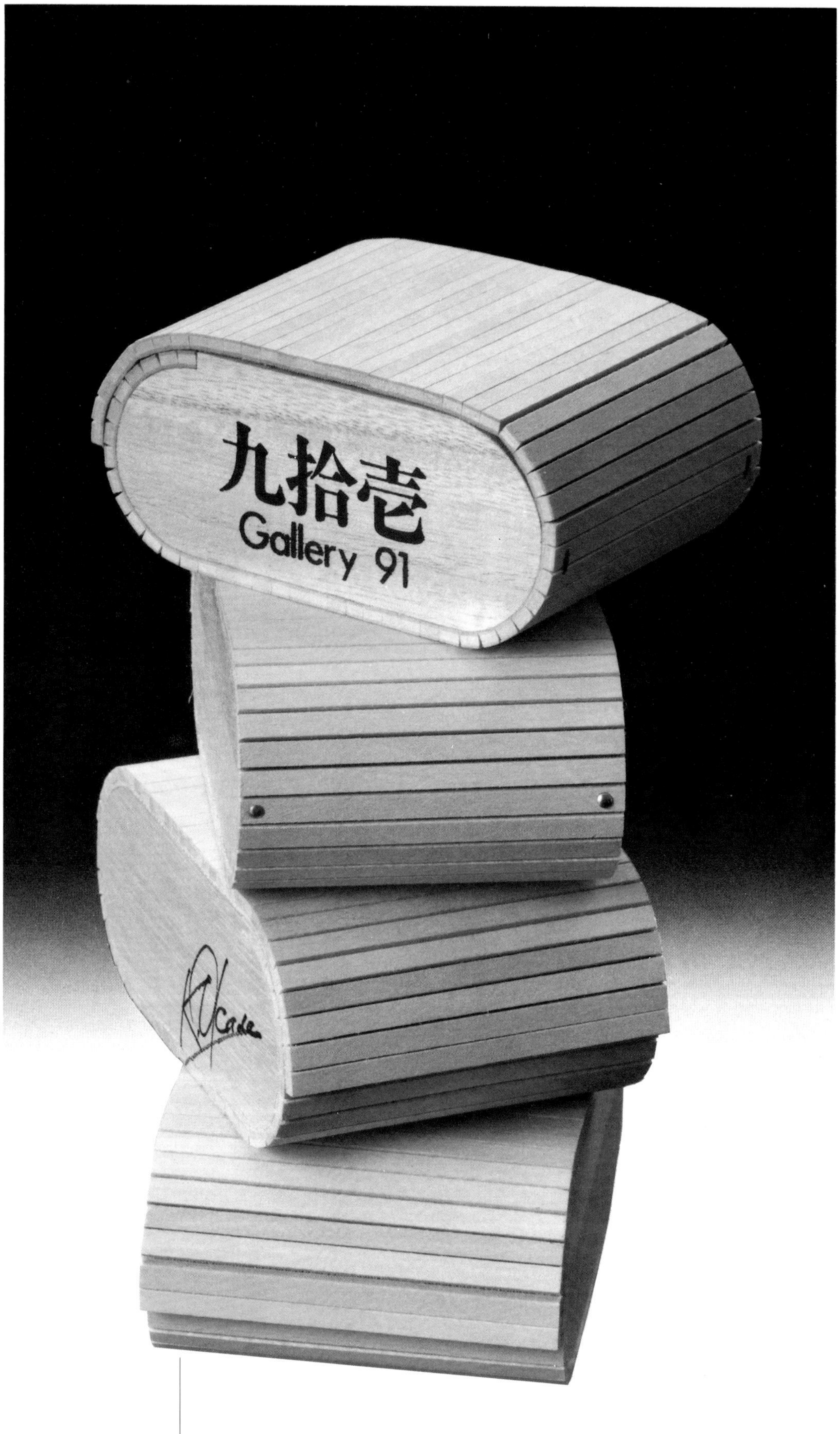

16

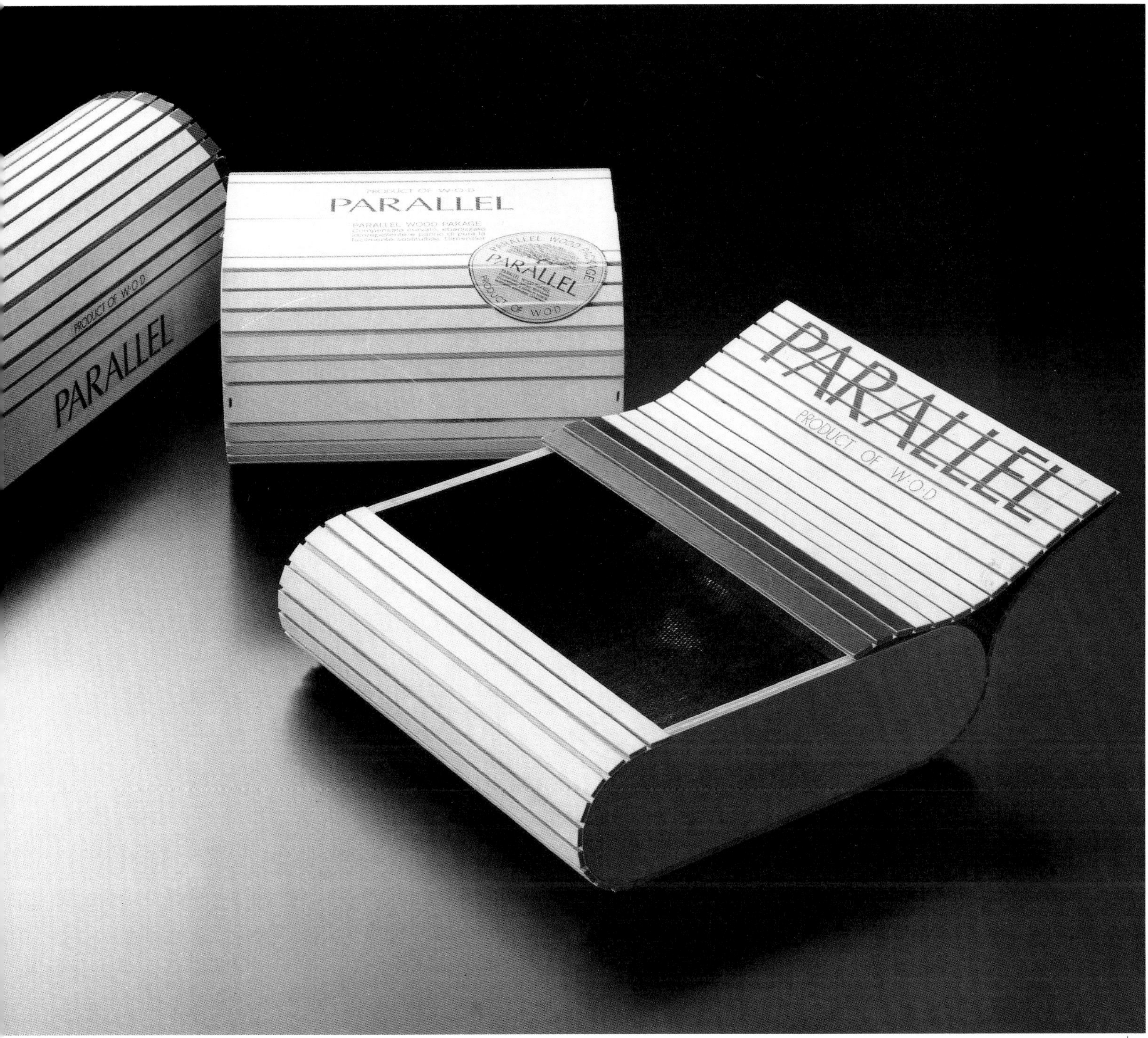
PRODUCT OF W·O·D
PARALLEL
PRODUCT OF W·O·D
PARALLEL
PARALLEL WOOD PAKAGE
PARALLEL
PRODUCT OF W·O·D
PARALLEL
PRODUCT OF W·O·D

18.........Whisky Gift Box
materials used : Paulownia Plywood & Cotton Cord
surface finish : Stripe Wood
client : The Nikka Whisky Distilling

19.........Whisky Gift Box
materials used : Paulownia Plywood, Paulownia & Cotton Cord
surface finish : Stripe Wood & Silk Screen Process
client : The Nikka Whisky Distilling

20.........Spice Case
materials used : Linden Plywood & Synthetic Paper
surface finish : Parallel Wood & Silk Screen Process

KOZO OKADA
岡田宏三

18……ウイスキー・ギフト・ボックス
使用素材……桐合板・綿ひも
表面処理……ストライプウッド
クライアント……ニッカウヰスキー

19……ウイスキー・ギフト・ボックス
使用素材……桐合板・桐・綿ひも
表面処理……ストライプウッド・シルクスクリーン印刷
クライアント……ニッカウヰスキー

20……スパイス・セット・ボックス
使用素材……シナ合板・特殊合紙
表面処理……パラレルウッド・シルクスクリーン印刷

18

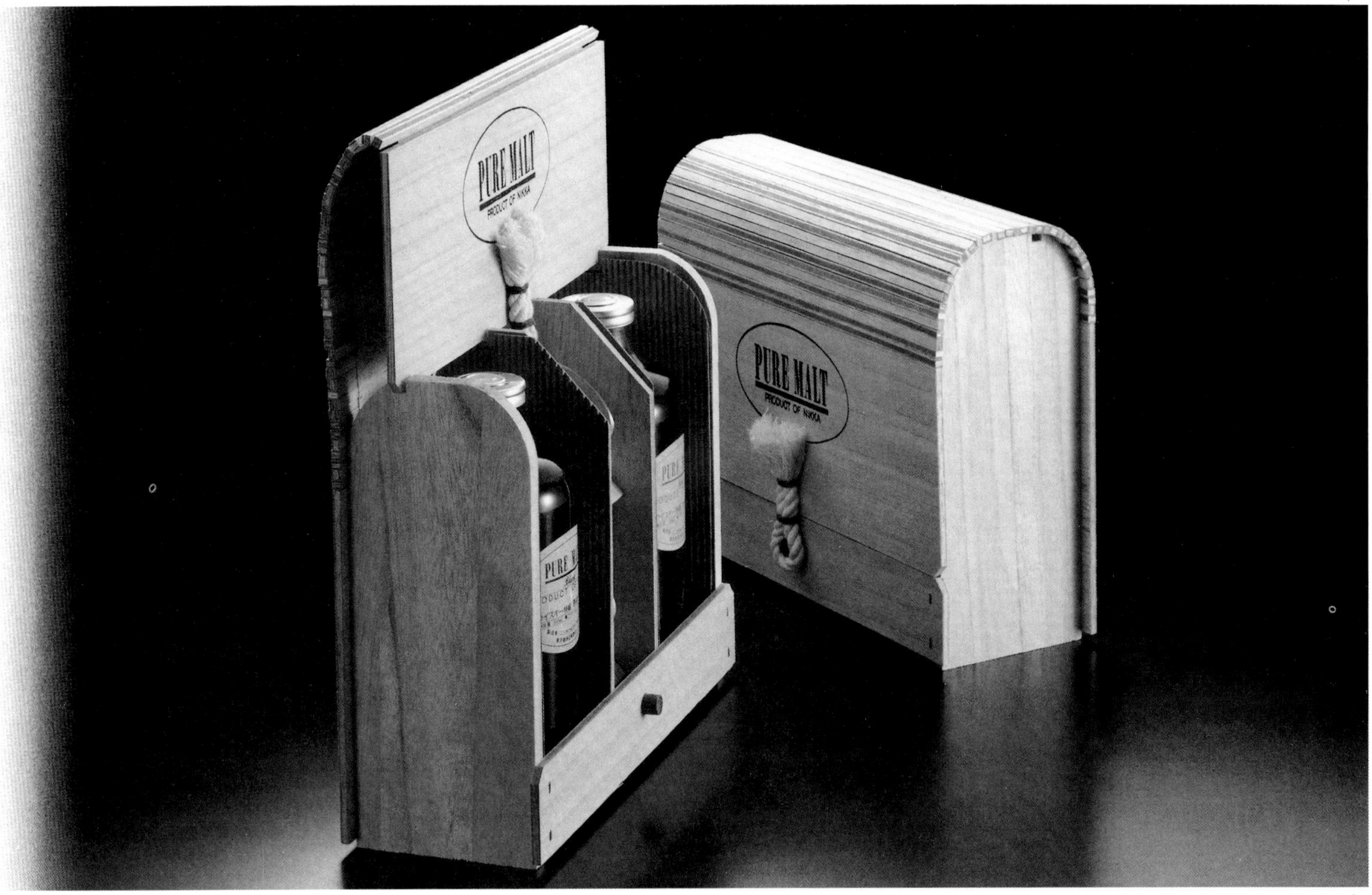
PURE MALT
PRODUCT OF NIKKA
PURE MALT
PRODUCT OF NIKKA

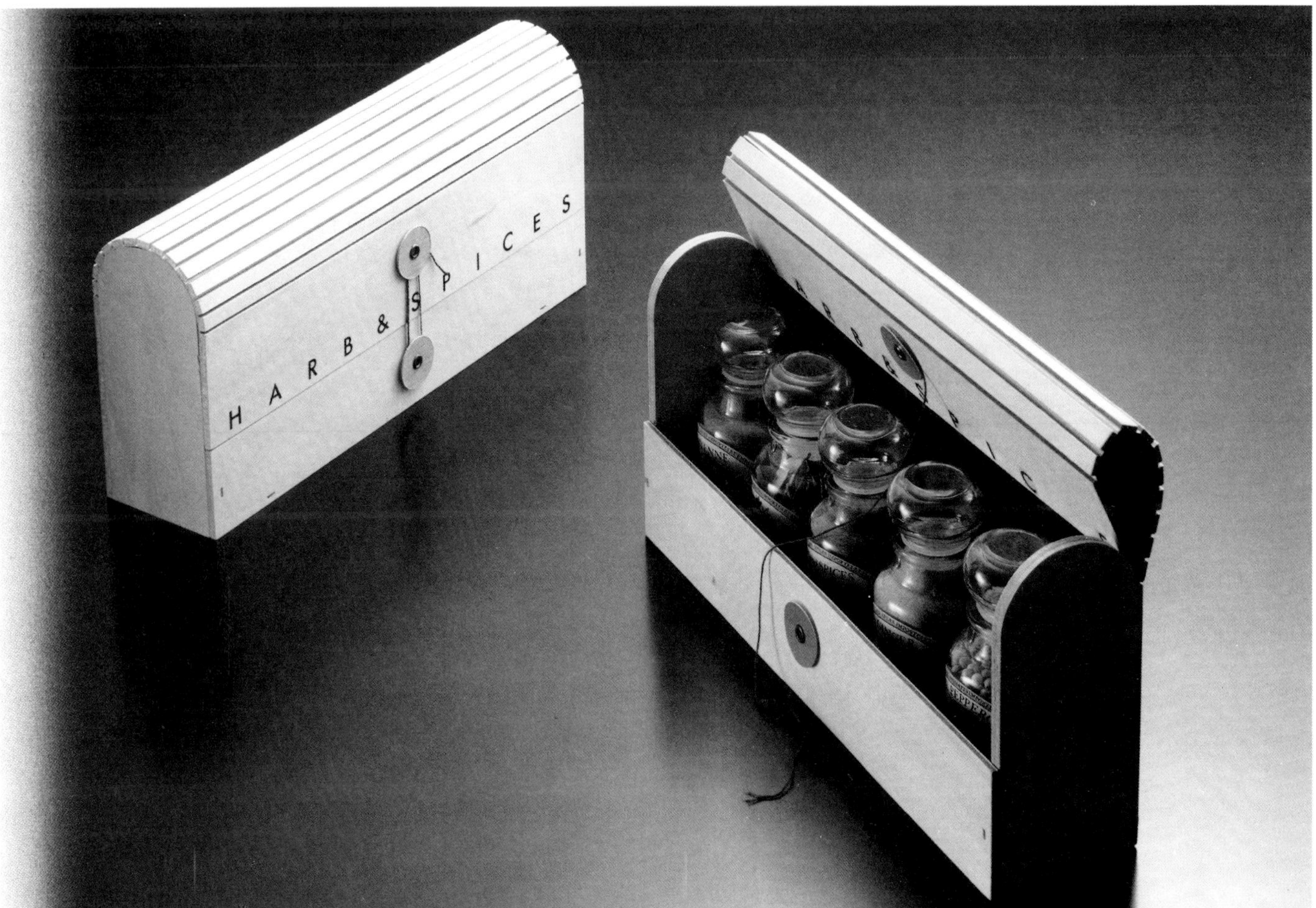
HARB & SPICES

21 **Wine Case**
materials used : Poulownia Plywood, Paulownia Cotton-linen & Cotton Cord
surface finish : Stripe Wood

22 **Flexible Box**
materials used : Paulownia Plywood, Paulownia & Cotton-linen
surface finish : Stripe Wood

23-24 **Boiled Fish Paste Container**
materials used : Paulownia Plywood, Paulownia & Cotton-linen
surface finish : Stripe Wood
client : Daitora

KOZO OKADA
岡田宏三

21……パッケージ・ヌーボー
使用素材……桐合板・桐・綿麻布・綿ひも
表面処理……ストライプウッド

22……フレキシブル・ボックス
使用素材……桐合板・桐・綿麻布
表面処理……ストライプウッド

23-24……かまぼこパッケージ
使用素材……桐合板・桐・綿麻布
表面処理……ストライプウッド
クライアント……大寅

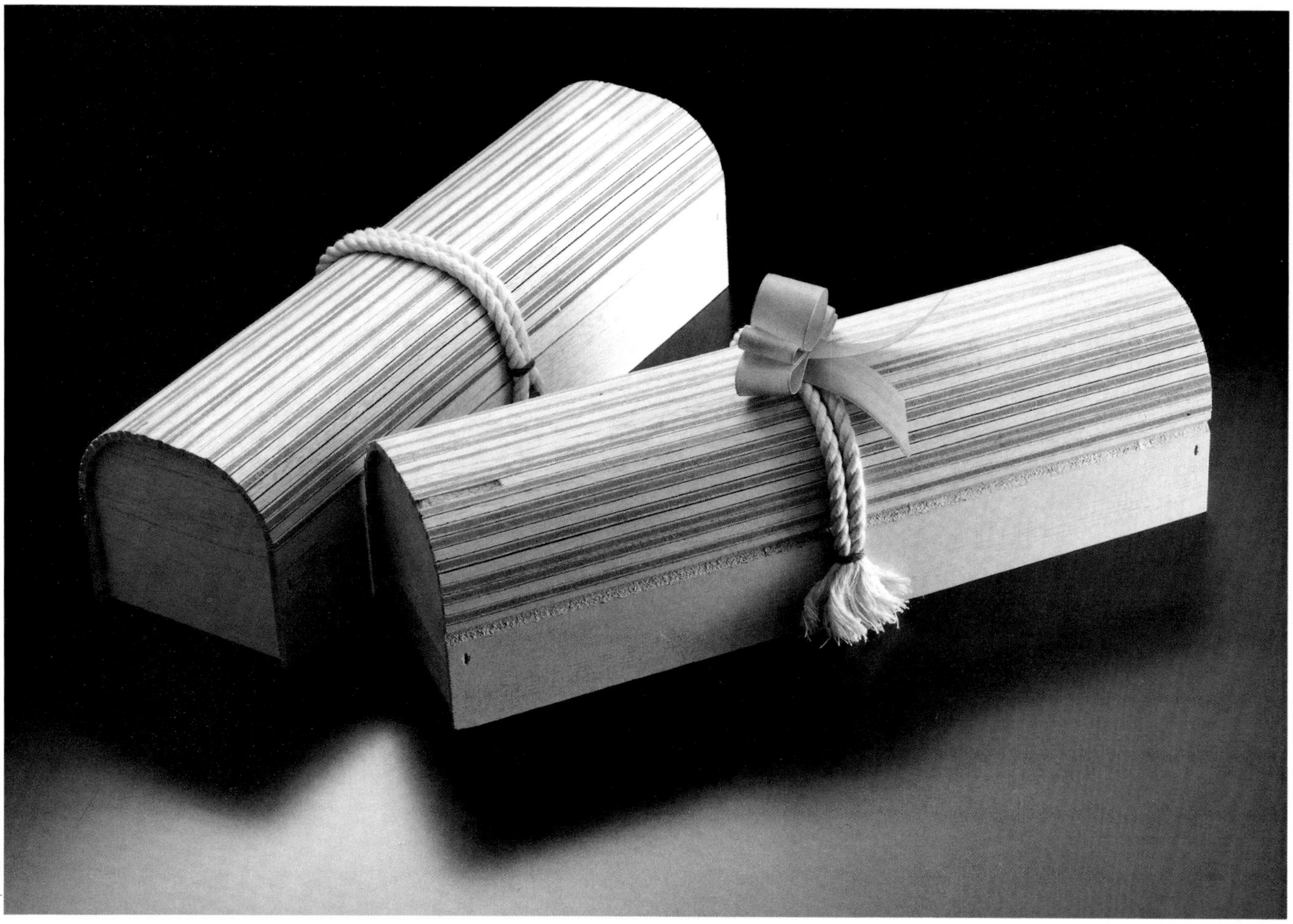
21

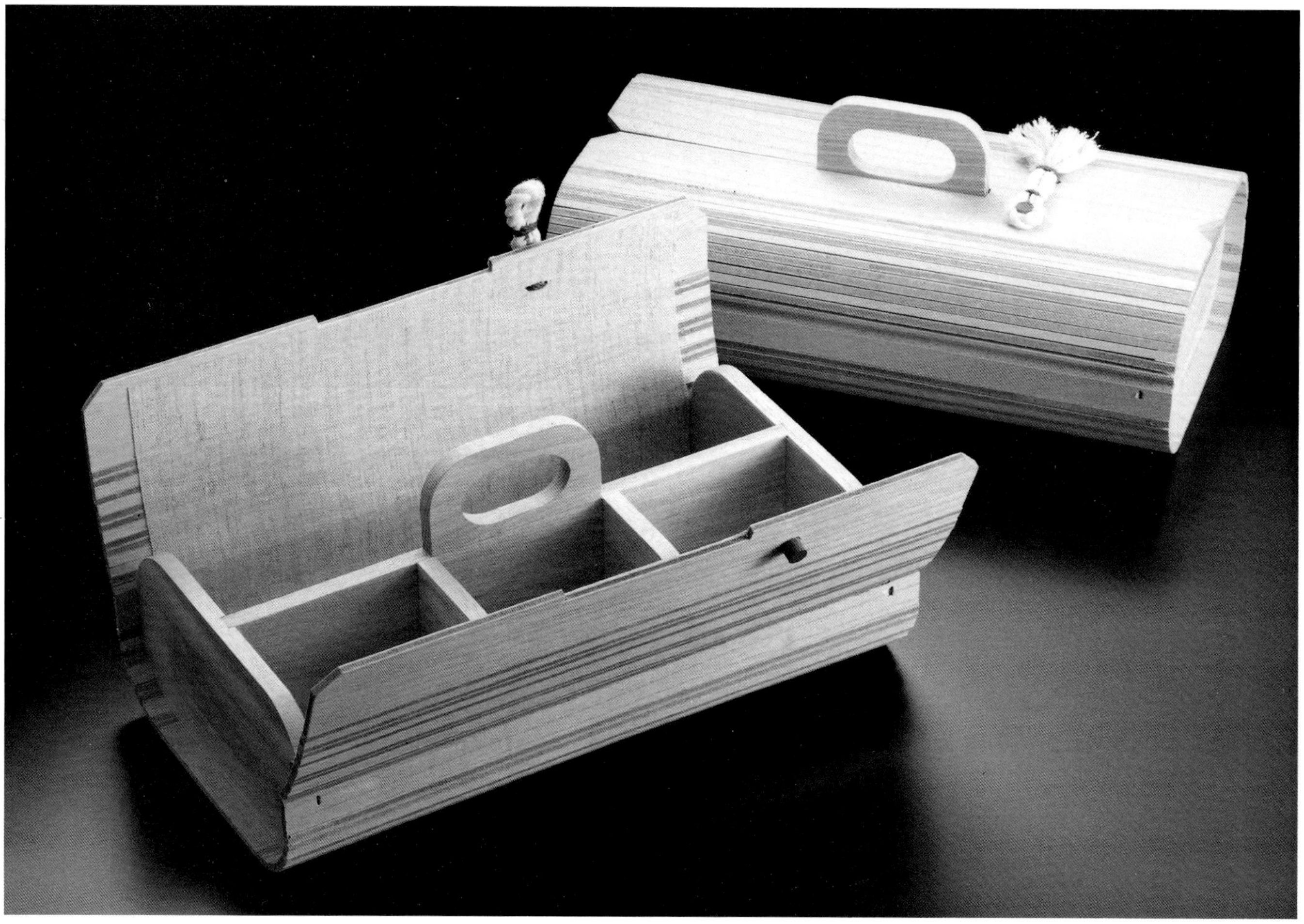
22

23

24

25............Whisky Bottle Cabinet
materials used : Linden Plywood & Tamo
surface finish : Clear Lacquer

26-27............Decoration Box
materials used : Linden Plywood & Synthetic Paper
surface finish : Parrallel Wood & Lacquer
client : Atom Chemical Paint

28............Tea Gift Package
materials used : Linden Plywood & Polyvinyl Chlorid Resin
client : Chakato

29............Tea Gift Package
materials used : Linden Plywood
surface finish : Silk Screen Process & Lacquer
client : Chakato

KOZO OKADA
岡田宏三

25……ウィスキー・キャビネット・ボックス
使用素材……シナ合板・タモ
表面処理……クリアラッカー塗装

26-27……自分デコレーション・ボックス
使用素材……シナ合板・特殊合紙
表面処理……パラレルウッド・ラッカー塗装
クライアント…アトム化学塗料㈱

28……お茶ギフト・パッケージ
使用素材……シナ合板・塩化ビニール
クライアント……茶加藤

29……お茶ギフト・パッケージ
使用素材……シナ合板
表面処理……ラッカー塗装・シルクスクリーン印刷
クライアント……茶加藤

25

26

27

28

新井亮

RYO ARAI

1.........Irritating Box
materials used : Paulownia & Linden Plywood
surface finish : Acrylic Resin

2.........Stacking Box
materials used : Paulownia & Elm Tree
surface finish : Acrylic Resin

RYO ARAI
新井 亮

1.........イライラ・ボックス
使用素材.........桐・シナ合板
表面処理.........アクリル樹脂塗装

2.........スタッキング・ボックス
使用素材.........桐・ニレ
表面処理.........アクリル樹脂塗装

1

3.......... Wall Pocket
material used : Beech Wood
surface finish : Silk Screen Process

4.......... Candy Box
material used : Linden Plywood
surface finish : Silk Screen Process

5.......... Letter Writing Kit Package
material used : Linden Plywood
surface finish : Silk Screen Process

RYO ARAI
新井 亮

3……ウォール・ポケット
使用素材……ブナ
表面処理……シルクスクリーン印刷

4……キャンディー・ボックス
使用素材……シナ合板
表面処理……シルクスクリーン印刷

5……レター・セット
使用素材……シナ合板
表面処理……シルクスクリーン印刷

3

4

CANDY BOX
CANDY BOX
CANDY BOX
CANDY BOX
CANDY BOX

5

LETTER SET
VIA AIR MAIL
Gift set
LETTER SET

6……Knockdown Gift Box
material used : Linden Plywood

7……Knockdown Gift Box
materials used : Linden Plywood & Beech Wood
surface finish : Silk Screen Process & Coloring

RYO ARAI
新井 亮

6……ノックダウン方式ギフト・ボックス
使用素材……シナ合板

7……ノックダウン方式ギフト・ボックス
使用素材……シナ合板・ブナ
表面処理……シルクスクリーン印刷・染料仕上げ

6

8.............Gift Box
materials used : Beech Wood & Leather
surface finish : Mat Urethane

9.............Gift Box coordinated Products of Saitama Pref.
material used : Paulownia

10.............Wear Box
materials used : Paulownia & Cloth Belt
surface finish : Silk Screen Process

RYO ARAI
新井亮

8.............**ギフト・ボックス**
使用素材.............ブナ・皮革
表面処理.............オープンボア仕上げ

9.............**ギフト・ボックス**
使用素材.............桐
埼玉県の物産をコーディネイト

10.............**ウェアー・ボックス**
使用素材.............桐・布ベルト
表面処理.............シルクスクリーン印刷

8

9

10

11……Boat-shaped Tray
material used : Small Scale Paulownia Plywood

12……Knockdown Gift Box
material used : Linden Plywood

RYO ARAI
新井 亮

11……舟形トレイ
使用素材……桐合紙

12……ノックダウン方式ギフト・ボックス
使用素材……シナ合板

11

RYO ARAI
新井 亮

13……ノックダウン方式ボトル・ラック
使用素材……杉シュロ縄

14……ノックダウン方式ワイン・ラック
使用素材……シナ合板

13

1984
Château
GRAVE
A PARGADE propriétaire à ILLATS
MIS EN BOUTEILLES AU CHATEAU

15……Box for Men
material used : Katsura
surface finish : Urethane

16……Magazine File Box
material used : Elm Tree
surface finish : Urethane

RYO ARAI
新井 亮

15……男のためのボックス
使用素材……桂
表面処理……ウレタン塗装

16……雑誌整理ボックス
使用素材……ニレ
表面処理……ウレタン塗装

15

cooking
fashion
make up · hair
culture

新井明子

AKIKO ARAI

AKIKO ARAI
新井明子

1……私に贈る私の箱・花・石庭・アブストラクト・海の幸
使用素材……桐
装飾……皮革・麻・造花用素材・合成皮革・ビーズ・塩ヒクロス・ロープ・石・ネット・ちりめん・ウールなど

1……Decorated Casket・Flower・Rock Garden・Abstract・Marine Products
material used : Paulownia
decoration : Leather, Linen, Materials for Artificial Flower, Synthetic Leather, Beads, Vinyl Chlorid Cloth, Rope, Stone, Net, Silk Crape, Wool etc.

1

2.............Gift Box
material used : Paulownia
surface finish : Gold Leaf Press

3.............Accessory Casket
material used : Paulownia
surface finish : Gold Leaf Press

4.............Dressing Table
material used : Paulownia
surface finish : Silk Screen Process

AKIKO ARAI
新井明子

2……サイコロのギフト・ボックス
使用素材……桐
表面処理……ホットスタンプ

3……アクセサリー箱
使用素材……桐
表面処理……ホットスタンプ

4……ポータブル鏡台
使用素材……桐
表面処理……シルクスクリーン印刷

2

3

4

5............Decorated Casket•Vegetable

material used : Paulownia
decoration : Silk, Wool, Cotton & Sea Weed etc.

AKIKO ARAI
新井明子

5............私に贈る私の箱•野菜

使用素材............桐
装飾............絹・ウール・綿・海草など

5

6……Memo Box
material used : Paulownia
surface finish : Silk Screen Process

7……Memo Pad
materials used : Linden Plywood, Cotton Cloth & Paper
surface finish : Silk Screen Process

8……Apparel Chest
material used : Paulownia
surface finish : Silk Screen Process

AKIKO ARAI
新井明子

6……メモ・ボックス
使用素材……桐
表面処理……シルクスクリーン印刷

7……メモ・パッド
使用素材……シナ合板・綿プリント生地・紙
表面処理……シルクスクリーン印刷

8……衣裳箱
使用素材……桐
表面処理……シルクスクリーン印刷

6

7

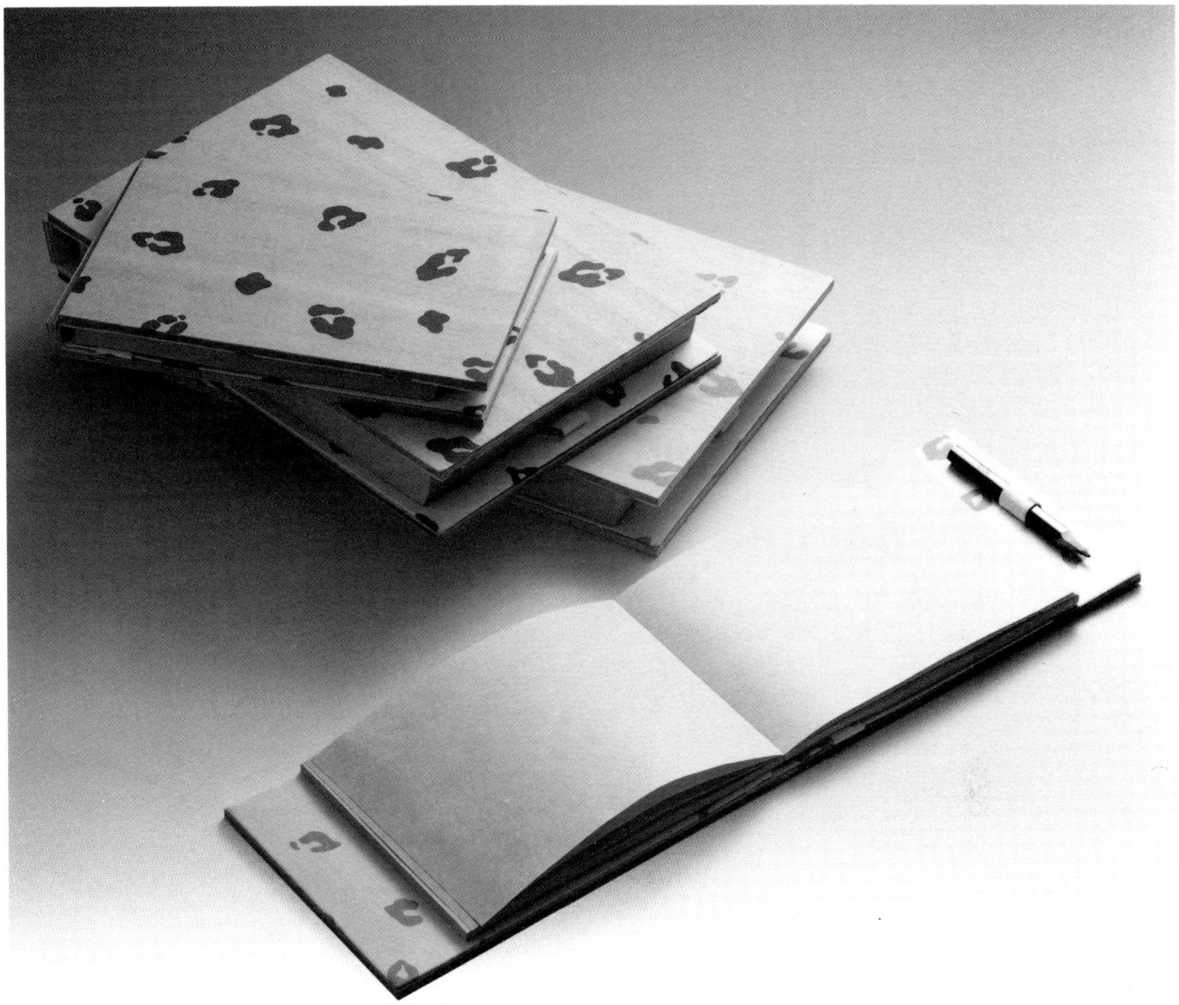

8

9 Cutlery Gift Case
material used : Paulownia
surface finish : Gold Leaf Press

10 Nest Casket
material used : Paulownia
surface finish : Silk Screen Process

AKIKO ARAI
新井明子

9 ギフト用カトラリー・ケース
使用素材 桐
表面処理 ホットスタンプ

10 入れ子の玉手箱
使用素材 桐
表面処理 シルクスクリーン印刷

9

11.......... **Bellows Box**
materials used : Linden Plywood, Paper Bellows & Ribbon
surface finish : Silk Screen Process

12.......... **Knockdown Memo Stand**
material used : Linden Plywood
surface finish : Silk Screen Process

13.......... **Letter File**
material used : Linden Plywood
surface finish : Silk Screen Process

14.......... **Acos Gift Package**
materials used : Linden Plywood & Cotton Cloth
surface finish : Silk Screen Process

15.......... **Acos Gift Package**
material used : Linden Plywood
surface finish : Silk Screen Process

AKIKO ARAI
新井明子

11……**ジャバラボックス**
使用素材……シナ合板・紙ジャバラ・リボン
表面処理……シルクスクリーン印刷

12……**分解式メモ・スタンド**
使用素材……シナ合板
表面処理……シルクスクリーン印刷

13……**レター・ファイル**
使用素材……シナ合板
表面処理……シルクスクリーン印刷

14……**アコス・ギフト・パッケージ**
使用素材……シナ合板・綿プリント生地
表面処理……シルクスクリーン印刷

15……**アコス・ギフト・パッケージ**
使用素材……シナ合板
表面処理……シルクスクリーン印刷

12

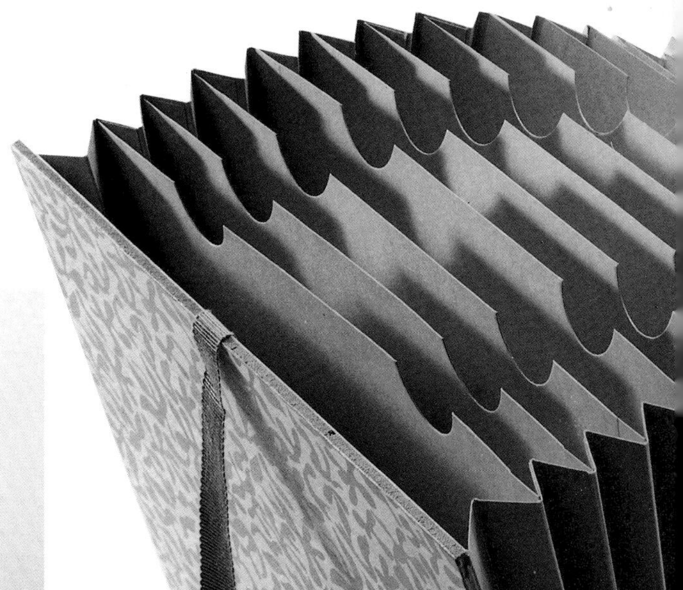

13

11

14

15

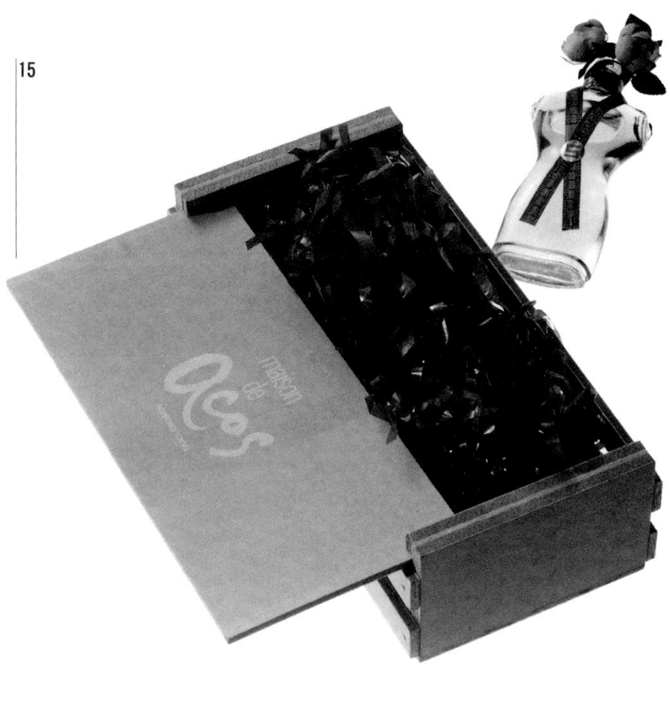

16.........Wrapping Cloth Box
materials used : Paulownia, Balsa & Fabric

17.........Knockdown Box
materials used : Plywood, Sliced Wood Sheet & Ribbon

AKIKO ARAI
新井明子

16……ふろしきボックス
使用素材……桐・バルサ・ファブリック

17……分解できる箱
使用素材……合板・木紙・リボン

16

秋月繁

SHIGERU AKIZUKI

1.............Box with Mask Illustration
materials used : Paulownia, Plywood & Cardboard
surface finish : Silk Screen Process & Appliqué

SHIGERU AKIZUKI
秋月 繁

1.............お面の箱
使用素材..........桐・合板・厚紙
表面処理..........シルクスクリーン印刷・切抜き・着色

1

2. Jack in the Box
material used : Pine
surface finish : Silk Screen Process

3. Chocolate Gift Box
material used : Paulownia
surface finish : Silk Screen Process & Coloration

4. Gift Box for Japanese Confectionery
material used : Paulownia
surface finish : Silk Screen Process

SHIGERU AKIZUKI
秋月 繁

2……びっくり箱
使用素材……松
表面処理……シルクスクリーン印刷

3……チョコレート・ギフト・ボックス
使用素材……桐
表面処理……シルクスクリーン印刷・着色

4……和菓子ギフト・ボックス
使用素材……桐
表面処理……シルクスクリーン印刷

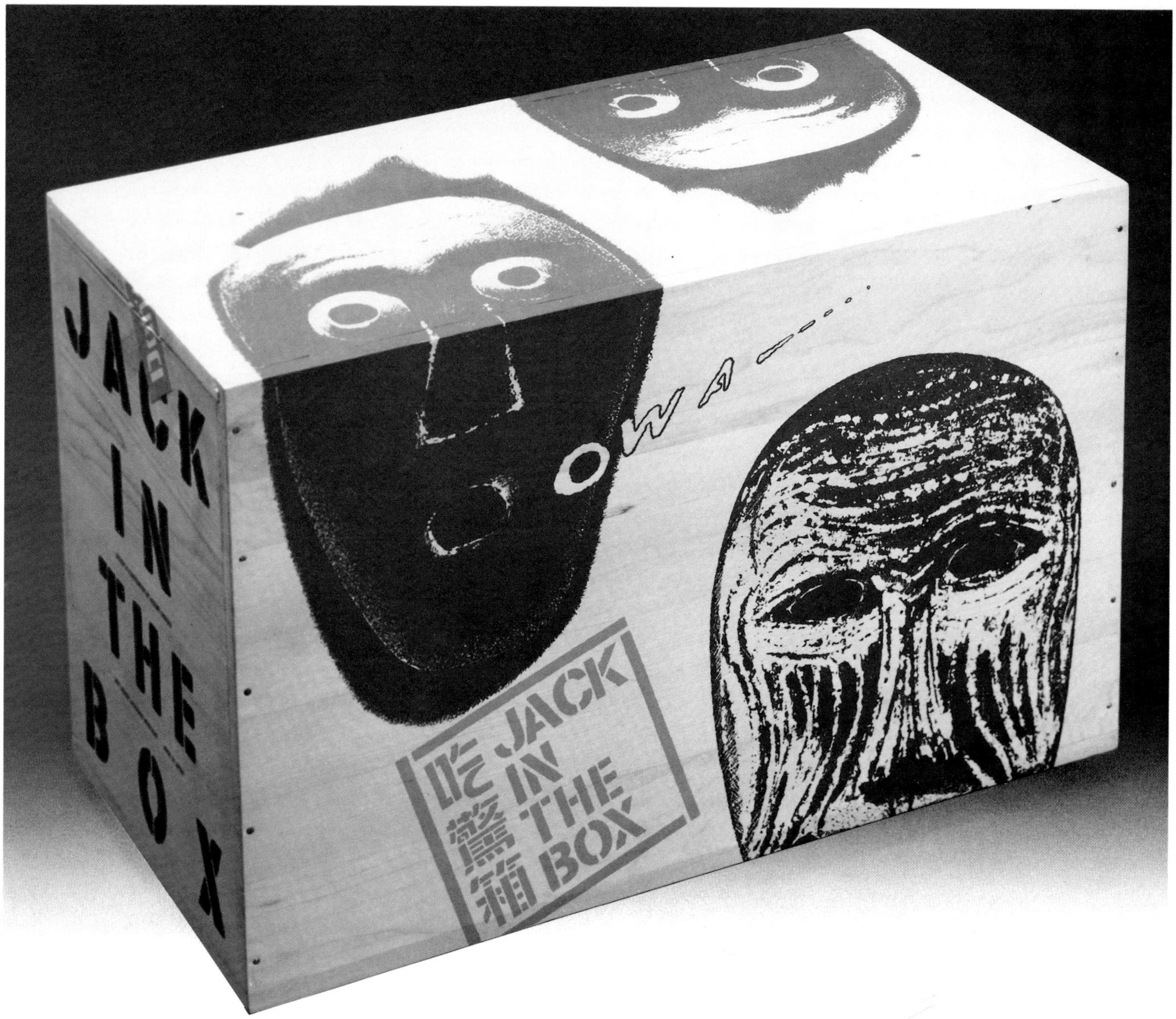

2

3

GIFT
BOX

5. Japanese Folk Mask 30
material used : Paulownia
surface finish : Silk Screen Process & Appliqué

6. Cassette Tape Case '78
material used : Linden Plywood
surface finish : Silk Screen Process & Appliqué

7. Sliding Eye Box
material used : Linden Plywood
surface finish : Silk Screen Process & Appliqué

SHIGERU AKIZUKI
秋月 繁

5……ジャパニーズ・フォーク・マスク30
使用素材……桐
表面処理……シルクスクリーン印刷・切抜き・着色

6……カセット・ボックス'78
使用素材……シナ合板
表面処理……シルクスクリーン印刷・切抜き・着色

7……スライディング・アイ・ボックス
使用素材……シナ合板
表面処理……シルクスクリーン印刷・切抜き・着色

5

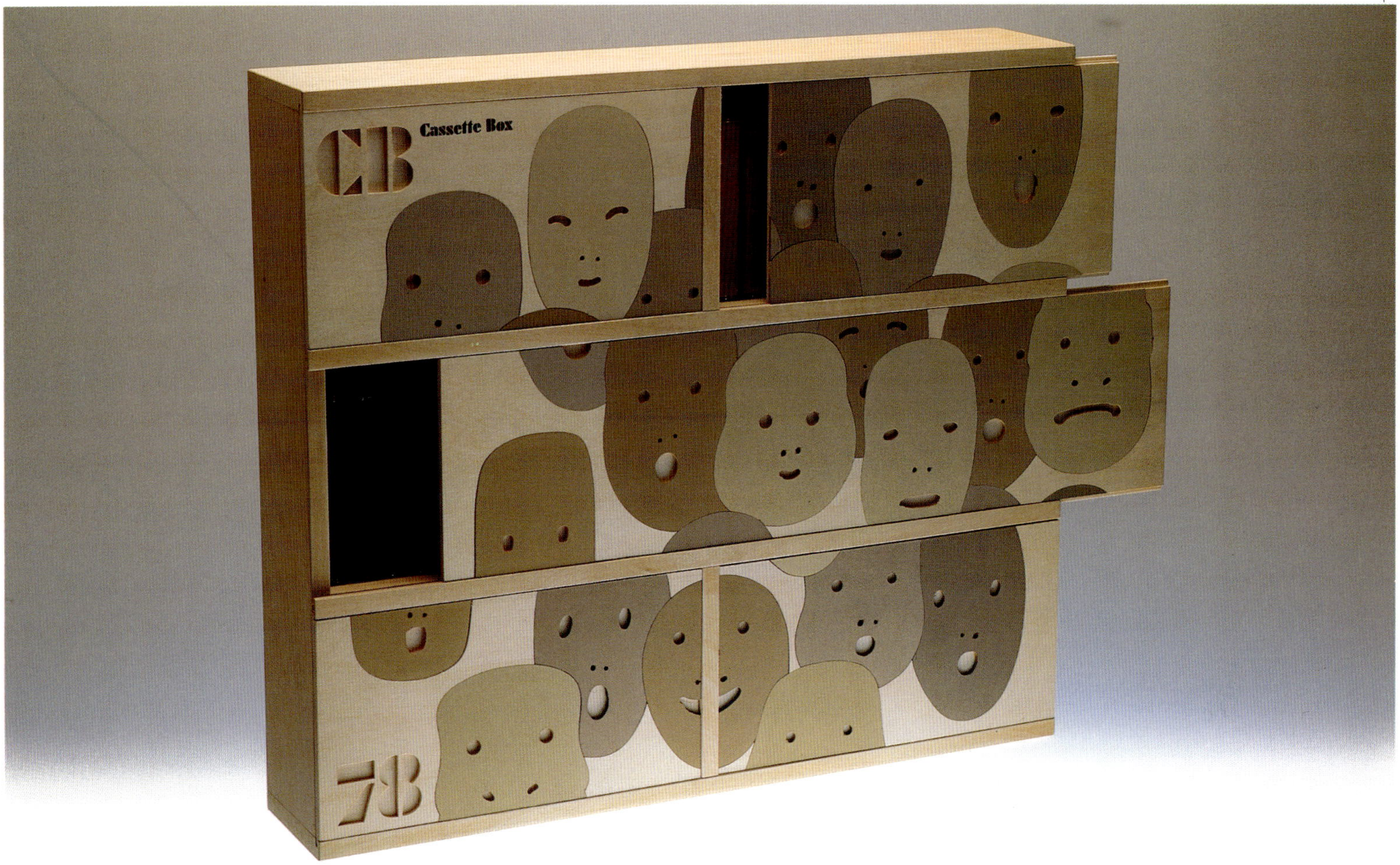
CB
Cassette Box
78

SLIDING
EYES
BOX

8.........Cassette Box '76
material used : Linden Plywood
surface finish : Silk Screen Process & Appliqué

9.........? Box
material used : Paulownia
surface finish : Appliqué

10.........Casket with Mask Illustration
material used : Paulownia
surface finish : Silk Screen Process

SHIGERU AKIZUKI
秋月 繁

8……カセット・ボックス'76
使用素材……シナ合板
表面処理……シルクスクリーン印刷・切抜き・着色

9……?箱
使用素材……桐
表面処理……切抜き・着色

10……お面の小箱
使用素材……桐
表面処理……シルクスクリーン印刷

8

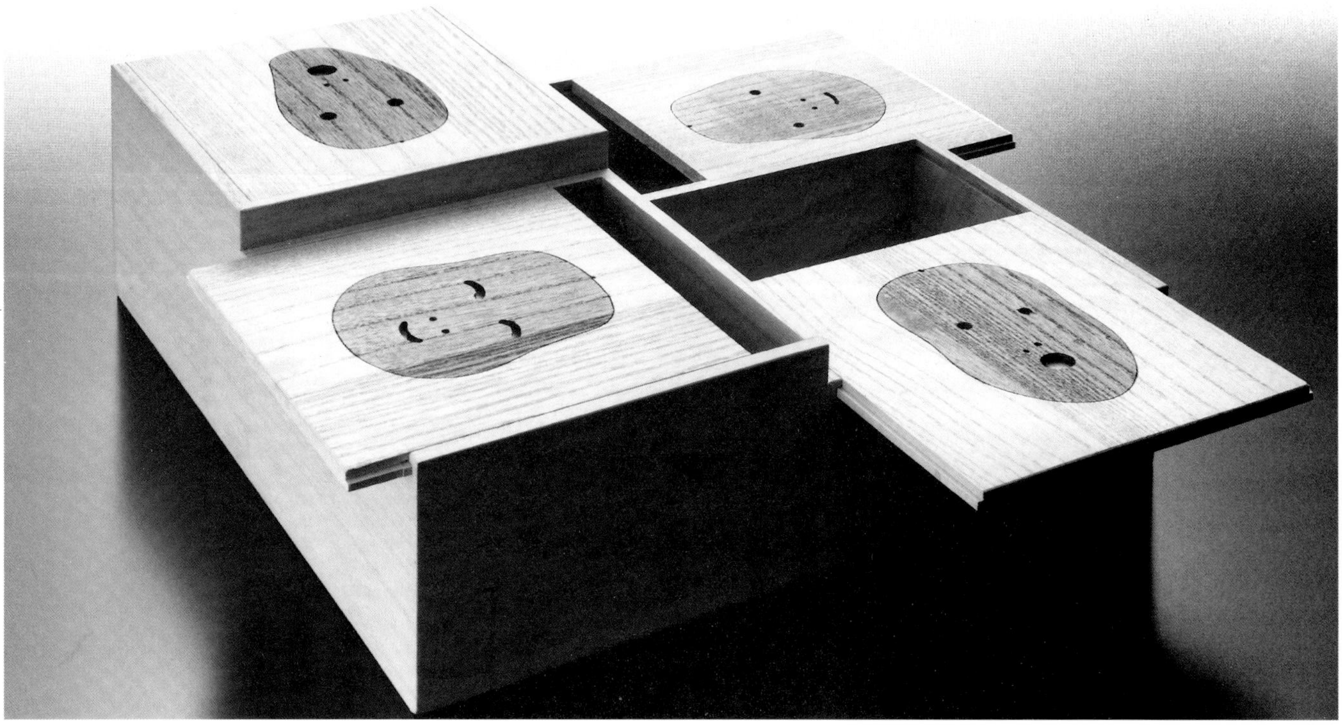

9

10

SHIGERU AKIZUKI
秋月繁

11.........文房宝箱
使用素材.........桐
表面処理.........切抜き·貼り合わせ

11

SHIGERU AKIZUKI
秋月繁

12……Gift Box for Whisky and Brandy
material used : Kalopanax
surface finish : Silk Screen Process & Appliqué

13……Wood Sample Box
material used :
surface finish : Silk Screen Process

14……Gift Box for Japanese Liquor
material used : Paulownia
surface finish : Silk Screen Process

12……ウイスキー・ブランデーのギフト・ボックス
使用素材……松
表面処理……シルクスクリーン印刷・切抜き・着色

13……木の見本板20
使用素材……センノキ
表面処理……シルクスクリーン印刷

14……日本酒のギフト・ボックス
使用素材……桐
表面処理……シルクスクリーン印刷

12

13

月下獨酌
大吟醸
江戸乃酒
DESIGN BY AKIZUKI SHIGERU
BOX MAKER : YAMADA KIRIBAKO SEISAKUSHO

material used : Paulownia
surface finish : Silk Screen Process

material used : Japanese Cypress
surface finish : Gouache & Waxing

SHIGERU AKIZUKI
秋月繁

15……ダルマ・ボックス
使用素材……桐
表面処理……シルクスクリーン印刷

16……ダルマの引出し
使用素材……桧
表面処理……グワッシュ・蝋付け

15

ヘブリン丸
中川晴壽堂
子ツ
サマシ
カゼヒキ
ノクスリ
消熱丸
有馬御湯花
本家
陳熟艾
中嶋屋新兵衛

17…………Chocolate Collection
materials used : Plywood & Cloth-paper
surface finish : Silk Screen Process

SHIGERU AKIZUKI
秋月 繁

17……チョコレート・コレクション
使用素材……合板・クロスペーパー
表面処理……シルクスクリーン印刷

17

18. Envelope
material used : Paulownia
surface finish : Silk Screen Process

19. Drawers
material used : Cherry Tree
surface finish : Coloration

SHIGERU AKIZUKI
秋月 繁

18 桐の封筒
使用素材……桐
表面処理……シルクスクリーン印刷

19 男の顔の引出し
使用素材……桜
表面処理……着色

18

20…………Wine Collection
material used : Pine
surface finish : Silk Screen Process

SHIGERU AKIZUKI
秋月繁

20……ワイン・コレクション
使用素材……松
表面処理……シルクスクリーン印刷

20

Profile

プロフィール

広橋桂子

●略歴

1931　東京生まれ
1954　東京芸術大学美術学部図案科卒業、森永製菓入社
1962　森永製菓退社、ヒロハシデザイン室主宰
1966-1976　東京芸術大学非常勤講師
1978　玉川大学デザイン研究室勤務、現在に至る。

●展覧会

1977　One Day One Show・東京デザイナーズスペース
1981　第1回4BOX-ER展・銀座和光
　　　遊びの木箱展・北海道近代美術館
1984　第1回ウッドパッケージ展・AXISギャラリー
　　　第2回4BOX-ER展・銀座和光
1986　朝日現代クラフト展委嘱出展
　　　第2回ウッドパッケージ展・AXISギャラリー

●受賞

1965　日本パッケージデザイン協会会員賞

現在、(社)日本パッケージデザイン協会会員
　　　(社)日本グラフィックデザイナー協会会員

◇木への回帰

技術革新という名のもとに、無機質工業化材料による商品群が氾濫している今日、
「木」への再評価は人間の感覚が求める必然的な道程だと思われます。
贈答用の包装にも自己表現を凝らしたり、贈り手の心が求められる時代に、
木肌の感触は気取りのないあたたかさとして受け入れられてきたように思われます。
木肌の魅力、白木の清浄感、桐の素材を十分に生かしながら種々の詰合せに適応できるノックダウン形式のギフトボックス。
平滑な木肌とジグザグ木肌の取り合わせ、型抜きなどヴィジュアルに楽しませる白木シリーズ。
これ等と全く対照的に全面塗装し、パターンをつけ曲面を生かした形態のパーソナルボックス。
「モノ」が生活に潤いを与えてくれたり、快適さを味あわせてくれたりということを願いながら制作してみました。

KEIKO HIROHASHI

1931 born in Tokyo
1954 graduated from Tokyo National University of Fine Arts and Music, Fine Arts Department, Design Section, and was employed by Morinaga & Co. Ltd.
1962 resigned Morinaga and started Hirohashi Design Room
1966-76 part-time lecturer at Tokyo National Univ. of Fine Arts and Music
1978-to date employed by Design Research Unit of Tamagawa University Design Research Center

Exhibition

1977 "One Day One Show", Tokyo Designers' Space
1981 "4 Box-ers" exhibition, Ginza Wako Gallery
"Playing Wooden Boxes"
exhibition, Hokkaido Prefectural Museum of Modern Art
1984 The 1st Wood Package Exhibition, AXIS Gallery, Tokyo
The 2nd "4 Box-ers" exhibition, Ginza Wako Gallery
1986 "Asahi Contemporary Crafts Exhibition '86"
The 2nd Wood Package" Exhibition, AXIS Gallery, Tokyo

Award

1965 Member's Award of the Japan Package Design Association Competition

Membership

Japan Package Design Association
Japan Graphic Designers Association

Wood as an Antithesis to Hi-tech Materials

It is but natural for human beings to re-evaluate "wood" amid abundant merchandise of inorganic, high-tech materials.
Today, originality is sought for even in wrapping a gift to convey the feelings of the sender.
In such a time, the touch of wood seems to be widely accepted as giving a plain and warm impression.
Presented here are knockdown gift boxes,
plain wood series and "personal boxes".
The texture and purity of paulownia are expressed for knockdown gift boxes
which can be flexibly used as gift containers.
The boxes of plain wood series combined with plain and rough grains,
and cutouts, could attract the viewers' eye.
In contrast, "personal boxes" are painted and shaped differently making use of curves.
I designed these boxes with the expectation that it will give charm and amenity to the user's daily life.

杉村敏男

●略歴

1939　静岡県生まれ
1978　株式会社すぎ・デザイン研究所設立、現在に至る。
1986　東京都立高等訓練校講師

●展覧会

1984　第1回ウッドパッケージ展・AXISギャラリー
1985　KIBAKO展・ニューヨーク、ギャラリー91
1986　第2回ウッドパッケージ展・AXISギャラリー

●受賞

1969　日本パッケージデザイン協会展特賞
1972　カレンダー展通商産業大臣賞
　　　パッケージング展工業技術院長賞
1974-1977　パッケージング展特別賞、食品部門賞、包装紙部門賞
1980　日本パッケージデザイン協会ニューヨーク展銀賞、銅賞
1981　パッケージング展通商産業省生活産業局長賞、特別賞
1983　JPC展部門奨励賞

現在、(社)日本パッケージデザイン協会理事
　　　(社)日本グラフィックデザイナー協会会員

◇かわりなきもの

あわただしい世の中の動きを見ていると、感動したり、驚ろいたりする事ばかりである。
サイクルの早さに、いささか戸惑いを感じている。
ブリキのおもちゃ、戦争中の貯金箱、古いビンなどいくつか集めていて、
ひとつひとつ素朴で味があり、その時代の背景がある。
前年に出たパッケージがもうなくなって新しいパッケージに変っている。
思い出を残す余裕すら無い。
長い歴史のある会社がマークを変え、社名を変えてCI計画を取り入れているが、人間の意識の持ち方だと思う。
古いものには威厳を感じる。
コマーシャリズムの中で仕事をしていると新しさがデザインと感違いする事もある。
今後木に触れながら、釘を何回も打ち直しながら使っていける箱や引出しを作ってみたい。

TOSHIO SUGIMURA

1939 born in Shizuoka
1978 established Sugi Design Institute Inc.
1986 became a lecturer of Tokyo Metropolitan Technical High School

● Exhibition

1984 The 1st Wood Package Exhibition, AXIS Gallery, Tokyo
1985 "KIBAKO Exhibition", Gallery 91, New York
1986 The 2nd Wood Package Exhibition, AXIS Gallery, Tokyo

● Award

1969 First Prize, Japan Package Design Association
1972 Minister of International Trade and Industry's Prize (MITI), Calendar Contest
1972 Director's Prize, Agency of Industrial Science and Technology, Packaging Exhibition
1974-77
Special Prize, Prize for Food Category, Prize for Wrapping Paper Category, Packaging Exhibition
1980 Silver and Bronze prizes, Japan Package Design Association New York Exhibition
1981 Director's Prize, Consumer Goods Industries Bureau, MITI, Packaging Exhibition
1983 Prize for Encouragement, JPC Exhibition

● Membership

Japan Package Design Association (director)
Japan Graphic Designers Association

Re-Using Boxes

There are lots of surprises and impressive happenings in today's fast moving world.
I often feel flurried with the rapid pace of change.
Each piece of my collection of tin toys,
saving boxes and bottles made during and right after World War II
has a special naive and artless taste reflecting the time of production.
Some of the packages made in earlier years have disappeared but no one cares about it.
Companies with long-standing history change their names
and introduce a whole set of corporate identity concept in their publicity and packages.
In the world of commercialism, there is an illusion that novelty is the essence of design.
That is not right.
I personally find dignity in old things.
I would like to devote myself to touching wood, making boxes and drawers that can be repeatedly used.

U.G.サトー

●略歴

1935 東京に生れる。東京学芸大学中退、桑沢デザイン研究所卒業
本名:佐藤雄治

1975 デザインファーム設立、現在に至る。

1981 ラハチ・ポスター・ビエンナーレの国際審査員として、審査にたずさわる。

●展覧会

1972 「私の進化論」個展・ギャラリーオカベ

1976 「バランス感覚」個展・フマギャラリー

1977 「ネガティブ、ポジティブ」個展・フマギャラリー

1978 日本現代版画大賞展に出品

1980 「樹シリーズ」個展・フマギャラリー

1981 ユーモアと諷刺国際ビエンナーレ(ブルガリア)に出品
(ガブロボ'81)
第1回西武美術館版画大賞展出品
ラハチ美術館でセリグラフィ個展

1982 イビサグラフィック'82(スペイン)出品
視覚サーカス展出品・東京、松屋

1983 プリントギャラリー(アムステルダム)でセリグラフィ個展開催

1984 「U.G.サトーのナンセンス・ワールド」個展・IBM、
川崎市民ギャラリー
「フェンス・シリーズ」個展・ギャラリーオリーブ
第1回ウッドパッケージ展・AXISギャラリー
現代のユーモア展・埼玉近代美術館
現代日本ポスター展・the Gallery Stratford

1985 ゆらゆらオブジェ展・池袋西武、アトリエヌーボー
KIBAKO展・ニューヨーク、ギャラリー91
日米平和ポスター展・ひろしま美術館、東京松屋
レーザー・ビーム技術によるCag展・ギャラリースペース21
U.G.サトーのナンセンスイメージ展・阪急百貨店

1986 第2回ウッドパッケージ展・AXISギャラリー
JAGDA平和ポスター展・ひろしま美術館

●受賞

1977 日本グラフィック・デザイン・ニューヨーク展銀賞

1978 ブルーノ・グラフィックデザイン・ビエンナーレ(チェコスロバキア)金賞
ワルシャワ国際ポスター・ビエンナーレ(ポーランド)佳作賞

1979 ラハチ・ポスター・ビエンナーレ(フィンランド)金賞

●著書

思索ナンセンス選集3「U.G.サトーの進化論世界」思索社
「どうぶつえん」福音館書店
「2本足と4本足」(香原志勢と共著)福音館書店
「U.G.サトーとイラストレーション」美術出版社

◇閉じこめられた世界から、開かれた世界へ

包む、しまう、隠す、保護する、大切にする………。
箱には重いイメージがつきまとう。
さらに木の箱となれば、堅牢で重厚な因習のにおいがはなれにくい。
ぼくのへそまがりなバランス感覚は、そこからできる限り脱出を試み、異なった空間へ飛翔してみようと謀反をはかるのです。
そこで日頃、ぼくがイラストレーションの仕事で使いこなしてきた、ナンセンスなトロンプ・ルイユの方法を導入。
平面的な手口を立体的な思考のなかに交差させます。
このだましのテクニックで、少しでも驚きや笑いを起こせれば成功。
閉じこめられた重い世界が、開かれた軽みの世界に変ればうれしいのです。

U.G. SATO

1935 born in Tokyo, left Tokyo Gakugei University, graduated from Kuwazawa Design School
1975 established U.G. Sato & Design Farm
1981 Lahti Poster, Biennale as a Jury

Exhibition

one man show

1972 "My Evolutionary Theory" exhibition, Gallery, Okabe, Tokyo
1976 "My Sense of Balance", Fuma Gallery, Tokyo
1977 "Negative, Positive" exhibition, Fuma Gallery
1978 Japan Contemporary Prints Exhibition
1980 "Wood Series" exhibition, Fuma Gallery
1981 Serigraphy exhibition, Lahti Museum, Finland
1983 Serigraphy exhibition, Print Gallery, Amsterdam, Netherlands
1984 "U.G. Sato's Nonsense World", IBM Kawasaki Civic Gallery
"Fence Series" exhibition, Gallery Olive, Tokyo
1985 "U.G. Sato's Nonsense Image", Hankyu Department store group, international exhibition
1981 Gabrove '81, International Biennale of Humor and Satire, Bulgaria
The 1st Prints Exhibition, Seibu Museum
1982 Ipisa Graphic '82, Spain
Visual Circus Exhibition, Matsuya, Tokyo
1984 The 1st Wood Package Exhibition, AXIS Gallery, Tokyo
"Contemporary Humor" exhibition, Saitama Pref. Museum of Modern Art
1984 "Contemporary Japanese Poster" exhibition, Gallery Stratford
1985 "Swinging Objects" exhibition, Ikebukuro Seibu
"KIBAKO" exhibition, Gallery 91, New York
"US-Japan Peace Poster" exhibition, Hiroshima Museum and Matsuya, Tokyo
"Cag by Laser Beam" exhibition, Gallery Space 21, Tokyo
1986 The 2nd Wood Package Exhibition, AXIS Gallery, Tokyo
"JAGDA Peace Poster" exhibition, Hiroshima museum

Award

1977 Silver prize, Japan Graphic Design New York Exhibition
1978 Gold prize, Graphic Design Biennale, Brno, Czechoslovakia award, International Poster Biennale, Warsaw, Poland
1979 Gold prize, Lahti Poster Biennale, Finland

Publication

"U.G. Sato's Revolutionary World" Shisaku-sha
"Zoo" Fukuinkan
"Two legs and Four legs" (co-author) Fukuinkan
"U.G. Sato and Illustration" Bijutsu Shuppansha

From a Closed World to an Open World

Pack, tuck away, hide away, protect, treasure. . . . . . ,
a heavy image hangs on the box.
It is more so with the box made of wood for its solid and long-life image.
My crabbed sensitivity rises,
revolting to escape from conventional impression and to take flight into a different space.
Thus, I introduced the nonsense "trompe l'oeil" that I often use in my illustration works,
employing two-dimensional methods in three-dimensional thoughts.
It will be a success if this tricky technique can cause a little bit of surprise or even a smile among viewers.
I am pleased if the closed and heavy world turns into an open and light world.

岡田宏三

●略歴

1934　大阪生まれ
1953　大阪市都島工業建築科卒業
1958　京都市立美術大学(現、京都芸大)卒業、レナウン入社(本社宣伝課)
1965　岡田デザイン事務所設立
1969　岡田デザイン事務所社名変更法人化、株式会社OD設立
　　　現在に至るまで、グラフィックデザイン、パッケージデザイン、
　　　アドバタイジングデザインほかのデザイン活動。
　　　個人活動として、スペースデザイン、ブティックKENZO SHOP、レストラン、
　　　バーなどの設計、コミュニケーションサイン、オブジェ、毛筆アート、映像美術など。

●展覧会

1980　日本国際美術展(彫刻)
1984　第1回ウッドパッケージ展・AXISギャラリー
1985　KIBAKO展・ニューヨーク、ギャラリー91
　　　岡田宏三の木がるな「もてなしのカタチ」個展・銀座松屋
　　　TATAMI CHAIRと「もてなしのカタチ」個展・池袋西武
　　　四季と新しい「もてなしのカタチ」個展・ギャラリースペース21
　　　岡田宏三が提案する日本茶と新しい「もてなしのカタチ」個展・渋谷西武
　　　ジャパンクリエイティブ
　　　岡田宏三の視覚対話空間展ウヰスキーと「もてなしのカタチ」個展
　　　(ニッカウヰスキー後援)
　　　岡田宏三の「もてなしのカタチ」TV展覧会・TV東京「奥さまTVジョッキー」3月-6月
1986　第2回ウッドパッケージ展・AXISギャラリー
　　　岡田宏三「もてなしのカタチ」個展・Robinson's 春日部店
1986-1987　「初春のカタチ」個展・渋谷西武
　　　　　　「贈りのカタチ」個展・有楽町西武

●受賞

1961　パッケージ展入賞多数
1975　SDA賞銀賞、銅賞、他入賞
1976　日本パッケージデザイン協会展入賞
1980　ニューヨーク日本パッケージデザイン展銅賞
1982　カレンダー展通産大臣賞
1985　年鑑日本のパッケージデザイン奨励賞2点
1986　クリオアワード・クリオ賞(パッケージ)ほか

●その他の活動

(財)日本産業デザイン振興会・日本優秀デザイン商品開発指導(熊本県、岩手県、北海道)
春日部桐箱産地パッケージ開発指導
桐生和紙の新しい素材開発など、その他多くの産地指導

現在、日本サインデザイン協会会員
　(社)日本パッケージデザイン協会会員
　(社)日本グラフィックデザイナー協会会員
　POP広告協会会員

◆視覚対話の箱

ここに掲載される作品は、私が今取組んでいるデザインテーマにそった一連の仕事である。
このテーマは、一つのオブジェクトを持ちながら二つのサブジェクトが潜んでいる。
オブジェクトは、「視覚対話の世界」。
これをパッケージデザインに取入れた場合、「視覚対話のできるパッケージ」となる。
サブジェクトの一つは、「素材感の再発見」である。
それは、限定した素材感で見られがちな一定素材を見直し、新しい素材感を再発見しようという試みである。
もう一つのサブジェクトは、「もてなしのカタチ」である。
視覚対話空間を構成する場合、そのカタチの中に、もてなし心が潜んでいてほしい。
心地よく、しかも新鮮な素材感でパッケージを構成できれば、
そこには自づから「もてなし心」の内在するカタチになるのでは、と私はひとりで信じ込んでいる。
わかるウッドパッケージではなく、感じるウッドパッケージをめざして――

素材感別シリーズとして以下の展開がある。

①ストライプウッドパッケージ
　細い角材で質感の違うものをランダムに並べて裏打ちし、
　曲面利用を可能にしたパッケージ。
②カットウッドパッケージ
　木口面を小さく組み合わせて新しい素材感を求めたもの。
③スライスウッドパッケージ
　薄くスライスした質感の違う木を組み合わせ、
　ウッドパッケージとペーパーパッケージの中間的役割を持つパッケージとした。

KOZO OKADA

1934 born in Osaka
1953 finished Osaka Municipal Miyakojima Technical High School
1958 graduated from Kyoto City University of Arts and was employed by Renown Inc.
1965 established Okada Design Office
1969 reorganized the Office into OD Inc. and involved in graphic design, package design, and advertising design, etc, as a company, space design for boutiques and restaurants, and bars, communication signs, objects, brush arts and video arts as personal activities

Exhibition

One man show
"Form of Hospitality" exhibitions, Ginza Matsuya, Ikebukuro Seibu, Gallery Space 21, Shibuya Seibu, Robinson's Kasukabe
"Form of Hospitality" TV exhibition (March-June, 1986)
"Form of Early Spring" Exhibition (Shibuya Seibu)
"Form of Gift" (Yurakucho Seibu)

Group Exhibition

1980 Tokyo Biennale Exhibition (sculpture)
1984 The 1st Wood Package Exhibition, AXIS Gallery, Tokyo
1985 "KIBAKO Exhibition", Gallery 91, New York
1986 The 2nd Wood Package Exhibition, AXIS Gallery, Tokyo

Award

1961 Various prizes of Packaging Exhibition
1975 SDA (Sign Design Assn.) Silver, Bronze and other prizes
1976 award of Japan Package Design Association Competition
1980 Silver prize, Japanese Package Exhibition, New York
1982 Minister of International Trade and Industry's Prize, Calendar Contest
1985 Two Awards of Encouragement, Japan Package Design Annual Competition
1986 Clio Award (for package)

Other Activities

Instructor for the development of merchandise of design excellence promoted by the Japan Industrial Design Promotion Organization (Kumamoto, Iwate, Hokkaido Prefectures), for package development for paulownia, Kasukabe, Saitama Pref., for material development of hand-woven paper, Kiryu, Gunma Pref., etc.

Membership

Japan Sign Design Association
Japan Package Designers Association
Japan Graphic Design Association
POP Advertisement Association

Boxes for Visual Communication

The boxes appearing in this book were made along my recent design theme
of having one object with two hidden subjects.
The object is "the world of visual dialogue",
that is to design packages as pieces of communication with consumers.
One of the subjects is "re-discovering the texture of the material",
or to explore a new touch of the conventional packing materials which have been seen with limited impression.
The other subject is "the form of entertainment".
When a space of visual dialogue is to be composed,
an entertaining mind is needed.
I am confident that if a designer can enjoy designing a package with a fresh feeling toward the material,
a sense of entertaining mind comes out naturally.
Instead of designing wood packages that viewers can understand,
the theme of my package design is for the viewers to feel something.

My recent works can be categorized, material-wise, into the following :
1) Stripe Wood Package
Slender pieces of square timber with different texture are arranged at random and lined,
thus, enabling the use of curved face.
2) Cut Wood Package
Small headers are composed to create a new feel of material.
3) Sliced Wood Package
Wooden foils of different texture are composed to make a package
made of a material between wood and paper.

新井亮

●略歴

1933　栃木県生まれ
1957　東京芸術大学美術学部図案科卒業
1958　日本宣伝美術協会会員になる。
1959　(株)デスカ入社、河野鷹思氏に師事
1962　フリーランスのイラストレーター、デザイナーとして、日立家電、西武百貨店、国土計画、いすゞ自動車、IBM等の仕事に当る。
1975　銅版画を始める。

●展覧会

1978　版画個展・青山5610ギャラリー
　　　日動画廊版画グランプリ展
1979　日動画廊版画グランプリ展
1980　版画個展・銀座ミキモトギャラリー
1984　第1回ウッドパッケージ展・AXISギャラリー
1985　版画個展・京王百貨店
1986　第2回ウッドパッケージ展・AXISギャラリー

●受賞

1958　日本宣伝美術協会展奨励賞
1959　大日本印刷主催カレンダーコンテスト金賞
1960　毎日新聞社主催毎日新聞広告賞特別賞
1982　カレンダーコンテスト通産大臣賞

●著書

1986　絵本「花くいライオン」講談社刊

現在、(社)日本グラフィックデザイナー協会会員

Ryoh Arai

◇木箱の今日的発想

日本の文化は、木と紙の文化と言われ、我々の生活の中にも様々なかたちでかかわりを持って来た。
茶室の中での茶通箱などは我々日本人の美意識に深い感銘を与える。
又、桐ダンスの磨き抜かれた技術に与えられたその美しさも同様である。
この様な深く伝統に根ざした桐箱に挑戦する事は無謀に近く思われたが、
しかし今日的ライフスタイルという視点から発想することにより、
量産性、ローコストを目的とした組立式ボトルラックあるいはギフトボックス、
又は箱自体が独立した個性と用途を持つもの等種々のデザインが生まれた。
この様に思いつくままに制作したのがこれらの作品である。

RYO ARAI

1933 born in Tochigi Pref.
1957 graduated from Tokyo National University of Fine Arts and Music
1958 became a member of Japan Advertising Artist Club
1959 employed by Desca and worked under Takashi Kono
1962 became a freelance illustrator/designer, and worked for Hitachi, Seibu Dept. Store, Isuzu Motors, IBM etc.
1975 started etching

Exhibition

One man show
1978 Wood Block Prints Exhibition, Aoyama 5610 Gallery, Tokyo
1980 Wood Block Prints, Ginza Mikimoto Gallery, Tokyo
1985 Wood Block Prints, Keio Dept. Store, Tokyo Group exhibition
1978 Wood Block Prints Grand Prix Exhibition, Nichido Gallery, Tokyo
1979 same as above
1985 The 1st Wood Package Exhibition, AXIS Gallery, Tokyo
1986 The 2nd Wood Package Exhibition, AXIS Gallery

Award

1958 Award for Encouragement, Japan Advertising Artist Club Exhibition
1959 Gold Prize, Calendar Contest by Dai Nippon Printing, Co. Ltd
1960 Special prize, Mainichi Advertising Prize, Mainichi Shimbun

Arai ryo

1982 Minister of International Trade and Industry' prize, Calendar Contest

Publication

1983 Illustrated book "Flower eating lions" Kodansha

Membership

Japan Graphic Designers Association

Up-to-date Use of Paulownia

Japanese culture is said to be a culture of wood and paper.
In fact, both wood and paper are deeply associated with the life of the Japanese in various forms.
The box for cakes served at a tea ceremony house releases a profound appeal for Japanese aesthetic sensitivity.
The beauty of a masterpiece of paulownia chest of drawers is equally appealing
It seems too wild to tackle the long-established paulownia box,
but by having some design concept in up-to-date lifestyle,
knockdown bottle racks, and gift boxes and other boxes were introduced.
These can be mass produced,
retaining their own characteristics and uses.
The boxes in this book are products of casual ideas.

新井明子

●略歴

——— 東京に生まれる。
——— 女子美術大学芸術学部図案科卒業
——— 市田株式会社に入社、テキスタイルデザイナーとなる。
——— アドセンター株式会社に入社、ファッションディレクターとなる。
——— 株式会社泰和産業に入社、トータルコーディネーターとなる。
——— アコスファブリックハウスを設立、コスチュームファブリック、インテリアファブリック、ファッショングッズ、生活関連商品の企画デザインをし、製品を製造、メゾンドアコスのブランドネームで販売する。
——— ニューヨーク・アップルテクスポに出展、テキスタイル製品を販売する。
——— 通産省絹製品振興懇談会メンバーに委嘱される。
——— 株式会社大阪西川・三菱カロランと寝具ファブリックのデザインライセンスをメゾンドアコスのブランドで行う。
——— 株式会社中西儀兵衛商店とハンカチーフのデザインライセンス契約を行う。
——— 株式会社宮嶋と洋傘のデザインライセンスをメゾンドアコスのブランドで行う。

現在全てのライセンス契約を終る。
メゾンドアコスのブランドで全商品を直営店で販売。

Akiko Arai

◇束の間の衣(コロモ)

パッケージとは不思議なものだ。
どんなに美しくても、工夫がこらされていても最後には、とり捨てられる運命にある。
パッケージをデザインすることは、中味が人手に渡る迄の、束の間の衣(コロモ)を着せてやる作業なのだ。
桐という美しい素材と出逢った時、捨てさるパッケージの運命に、私はひどくこだわった。
本職のパッケージデザイナーと違って、パッケージのコマーシャルバリューと言うものに、今一歩踏み込むことが出来ない私は、
自分の今までの仕事、女としての特性をこの仕事で生かしたかった。
そこで、桐をモダナイズすることと捨てられない箱を考えてみたのである。
一回目より二回目に新しい発見がある。
この仕事はもっと追求してみたいテーマである。

AKIKO ARAI

——— born in Tokyo
——— graduated from the Women's College of Fine Art, Tokyo
——— employed by Ichida Inc. as a textile designer
——— employed by Ad Center Inc. and became a fashion director
——— employed by Taiwa Industry and became a total coordinator
——— established Acos Fabric House: costume fabric & interior fabric design, design of fashion goods and other daily life-related goods, sales of products under the brand name of Maison de Acos
——— participated in New York Apple Texpo and sold textile
——— appointed as a member of Silk Product Promotion Round Table, Ministry of International Trade and Industry
——— licensed the brand name Maison de Acos to the bedclothes produced by Osaka Nishikawa and Mitsubishi Carolan
——— licensed the brand name to handkerchief design of Nakanishi Gihei Shoten
——— licensed the brand name to umbrella design of Miyajima Co. Ltd.

At present, all the license contracts have terminated, and all the products of the brand name are sold at chain shops

Short-lived Clothes

A package has a pathetic life.
However beautiful it may be designed,
a package is destined to be thrown away.
The act of designing a package is putting a garment on an article for a brief period
until it is handed to a user.
When I came across the paulownia, a beautiful material,
I felt it is rather unfair that the life of a package is short.
Not being a professional package designer,
I can, free from seeking for commercial value,
venture my experiences and traits of a woman in designing paulownia packages.
Thus, I thought of a modern usage of paulownia and to design boxes
that will not be thrown away.
It is a new discovery for me with every box I make.
Paulownia is a material worth of pursuit.

秋月 繁

●略歴
1930　中国青島生まれ
1955　日本広告美術学校卒業、以来フリーデザイナーとして現在に至る。
　　　桑沢デザイン研究所講師

●展覧会
1976　版画個展・京王百貨店
1981　遊びの木箱展・北海道近代美術館
　　　第1回4BOX-ER展・銀座和光
1984　第2回4BOX-ER展・銀座和光
　　　第1回ウッドパッケージ展・AXISギャラリー
1986　第2回ウッドパッケージ展・AXISギャラリー
　　　日本のパッケージデザイン展・アメリカ、パーデュイ大学

●受賞
1967　第4回日本パッケージデザイン協会展会員賞
1969　第6回日本パッケージデザイン協会展会員賞
1973　第8回日本パッケージデザイン協会展会員賞
1985　年鑑日本のパッケージデザイン特賞、奨励賞

●著書
1975　「7人のパッケージデザイナー」六耀社刊
1982　「4人の箱」六耀社刊

●作品掲載
1985　グラフィス誌239号に個人特集
　　　その他国内、海外デザイン誌に掲載される。

現在、(社)日本グラフィックデザイナー協会会員
　　　日本郷土玩具の会会員

Shigeru Akizuki

◇個性豊かな素材

木には、なんとなく温もりがある。
手ざわりはカタイけど、なぜか柔らかく、それぞれ特有のニオイや、色ツヤを持っており個性ゆたかだ。
三年前に桧で作ったダルマの引出しなどは、今だに引出しを開けると桧の香りが楽しめる。
桐は、産地(春日部)特有の素材なので多く使ってみたが、
特長は軽いことでその軽さを活かして、桐の封筒を作ってみた。
定形郵便で50グラム迄(便せん3～4枚)70円でOKである。
それに、松、杉、合板などでダルマや土俗面の遊びのハコや、引出し。
又、日本酒、洋酒、菓子などのギフト用パッケージを作ってみた。
板を切り抜いたり、染めたり、シルク印刷したり、又、あるものは手描きだったりする。
こんな気(木)まいなハコやパッケージを、見て、さわって、なんとなく気持がやすらいでくれたら嬉しく思う

SHIGERU AKIZUKI

1930 born in Tsingtao, China
1955 graduated from Japan Advertising Art School, became a freelance designer
present lecturer, Kuwazawa Design School

Exhibition

1976 "Prints" One-man exhibition, Keio Dept. store
1981 "Playing Wooden Boxes" exhibition, Hokkaido Prefectural Museum of Modern Art
The 1st "4 Box-ers" exhibition, Ginza Wako Gallery, Tokyo
1984 The 2nd "4 Box-ers" exhibition, Ginza Wako Gallery
The 1st Wood Package Exhibition, AXIS Gallery, Tokyo
1986 The 2nd Wood Package Exhibition, AXIS Gallery
"Japanese Package Design" exhibition, Purdue University, USA

Award

1967 Member's award, the 4th Japan Package Design Association Competition
1969 Member's award, the 6th JPDA Competition
1973 Member's award, the 8th JPDA Competition
1985 Japan Package Design Award for Super Excellence, and Award for Encouragement

Publication

1975 "7 Package Designers in Japan", Rikuyo-sha Publishing, Inc.
1982 "Boxes by Four", Rikuyo-sha Publishing, Inc.

Featured Article

1985 Graphis No. 239 (Switzerland) and other design magazines both at home and abroad

Membership

Japan Graphic Designers Association
Japan Folk Toys' Society

The Relieving Touch of Wood

Wood has warmth. It is solid to touch but tender to look at.
Each tree has its own fragrance and luster and color, a kind of personality.
The drawers in the shape of a Bodhidharma doll which I made three years ago
still shed the fragrance of the Japanese cypress
when a drawer is drawn.
I made the letter envelope out of paulownia making use of its lightness.
Bodhidharma dolls, playful boxes of folk masks and drawers,
gift packages for cakes and liquor bottles. . . . .
these are all made of pine, cedar or plywood.
Wood is cut, painted, silk-screen processed or hand painted.
I am happy if customers feel relaxed when looking or touching these wood packages.

広告
ADVERTISEMENT

私はおいしいウイスキーを知っています。
NIKKA WHISKY

飲酒は20歳を過ぎてから

潜在力という言葉は、なまめかしい。
自然は生命力を育み、そこに潜在力を
ひそかに仕込む。北の風土と気候が熟成
させたモルト原酒の潜在力を、
あますところなく、めざめさせたい。
官能の導くままに、ニッカのブレンダー
たちは試みた。ここに、熟成モルト原酒が、
まず常識を越えた多量のレベルとなり、
かつ、多彩なモルトたちが互いに
潜在する香味を限界まで高め合う。そして
熟成カフェグレーンが、これら香味の
埋蔵量をさらに底の底までひきだす。
果て知れぬ香りをめざめさせた、この
ブレンドは、あなたの潜在する
官能をゆさぶるだろう。
「ザ・ブレンド・オブ・ニッカ」と名づける。

新発売

ザ・ブレンド・オブ・ニッカ

5,000円 660mℓ・alc45°・ウイスキー特級
価格は標準小売価格です。

モルトの、潜在力。

The Blend of Nikka

Maltbase Whisky

For Beautiful Human Life
Kanebo
唇にも年齢があります。
いま欲しいのは水分ですか、明るさですか。
〝しっとり水発色〟
初めてのアクアルージュがおこたえします。
アクアだから、この不思議。DUYアクアルージュの特長
❶初めての〝水のツヤ〟。若々しい明るさを輝かせます。
❷初めての〝水ののび〟。少量でもスムーズなのびが自慢です。
❸初めての〝水による保湿〟。バイオリキッドの配合で、唇にしっとり感を。
(マルチトール＋L-ヒアルロン酸)
あなたはご存知ですか、くちびるの生理学。
くちびるも皮膚の一部。年齢を重ねるにつれ肌と同じように張りやツヤを失なっていきます。またくちびるには汗腺がなく、皮脂の分泌が少ないため、自ら水分、脂分を補うことができません。ミセスのリップがかさつきやすくなるのはこのためもあるのです。
容器の構造
エピータ
デュイ アクアルージュ
全8色 各3,000円
唇にツヤと明るさ。不思議、しっとり、アクアルージュ。
DUY AQUA ROUGE
バイオリキッド配合(保湿)
100周年記念商品 第1弾
感謝をこめて、100周年
100 ANNIVERSARY
美しきヒューマンライフをめざす
カネボウ化粧品
★ベルの会へお入りください。
この看板のあるカネボウチェーン店・有名百貨店でお買いもとめください。
水のルージュ。
古手川 祐子

For Beautiful Human Life
Kanebo
メイクがくすむ…そろそろ「かわりめ肌」ですね。
古手川 祐子
感謝をこめて、100周年
100 ANNIVERSARY
この看板のあるカネボウチェーン店・有名百貨店でお買いもとめください。★ベルの会にお入りください。
美しきヒューマンライフをめざす
カネボウ化粧品
メイク映えも、透明感も以前と違う。それがかわりめ肌なんですね。
朝はちゃんとメイクがのっていたのに、午後になると、くすんだ感じになる。
透明感も以前と違うし、ときどきお肌もかさついたりする。
おかしいな…。メイク映えが気になりだしたら、そろそろかわりめ肌なんですね。
かわりめ肌と気づいたら、乳液を。ルネージュはバイオミルク成分配合。
かわりめ肌は、まず、水分と油分がバランスよく入っている乳液でのお手入れがポイントです。そこで、ルネージュのべっぴん乳液には、バイオミルク成分を配合しました。
お肌をしなやかにととのえるルネージュのべっぴん乳液。3つの特長。
①くすみがちなお肌に、うるおいとしなやかさを与えます。
②水分の蒸散をおさえ、お肌の乾燥を防ぎます。
③ミルクの感触で、お肌へのなじみがソフトです。
肌で、感触で、選べる3タイプ
さっぱりタイプの1番
やわらかで、さっぱりとしたタイプの乳液です。
さらっとタイプの2番
ソフトな感触とまろやかなコクのある乳液です。
しっとりタイプの3番
豊かなモイスチャー感のあるリッチなタイプの乳液です。
ルネージュ ミルキィローション
各3,000円
かわりめ肌に、バイオミルク成分配合
べっぴん乳液。
新発売 簡単スキンケア
エピータ
ルネージュ

ここは仕事場であり、いこいの場である。
ミラノがデザインするニットウエアの秘密は、
このウッディなスペースにあるようだ。

——U・G・サトー

写真は、建築界の鬼才・丸山欣也が設計した
ミラノのアビ・ルーム。
粋なパーティやパフォーマンスの場として、
どうぞお使いください。
お申込は下記のところまで……。

ニットを通じて自分発見！
milano group
株式会社ミラノ　新宿区市谷仲之町2-21　TEL03-358-8511(代)

ORIGINAL
We make planning designing everything for printing patterns.

PRINTS;
acos
Fabric House
株式会社 アコス ファブリック ハウス
代表　新 井 明 子
〒106 東京都港区六本木7-12-25
TEL (03) 403-4809
FAX (03) 403-4814

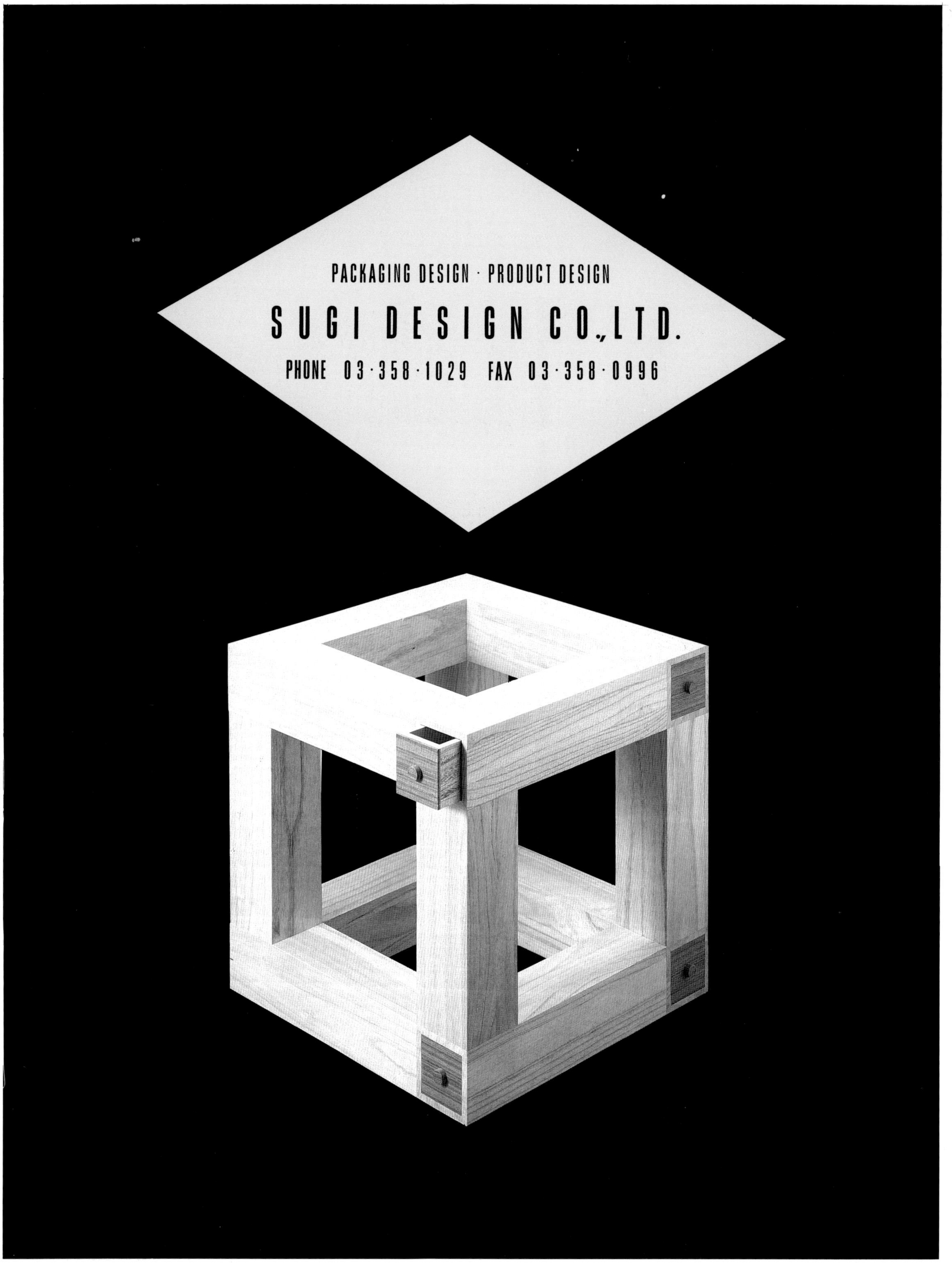
PACKAGING DESIGN · PRODUCT DESIGN
SUGI DESIGN CO.,LTD.
PHONE 03·358·1029 FAX 03·358·0996

化粧品からCDまで
パッケージのデモンストレーション

日本全国から公募した多量のパッケージの
中から、選び抜かれた価値ある作品だけを収録。
極めつきのパッケージデザイン年鑑。

年鑑 日本のパッケージデザイン1987

- ●内容　A部門：販売戦略的パッケージ
  　B部門：機能的パッケージ
  　C部門：デザイン意図的パッケージ
  　D部門：実験的試作品
  　掲載点数　450余点
- ●編集　(社)日本パッケージデザイン協会
- ●ブックデザイン／構成　勝井三雄
- ●体裁　菊４判(天地300×左右230㎜)
  　総360頁うちカラー頁247頁
  　角背上製本
- ●定価　16,000円

六耀社

木箱製作グループ

厚川木工所•厚川秀樹
鈴木木工所•鈴木行雄
松田桐箱製作所•松田行雄
山田桐箱製作所•山田隆夫
大野紙器製作所•大野雅司
春日部市桐箱工業協同組合

Wood Box Manufacturer's Group

Atuskawa Woodworking Shop•Hideki Atsukawa
Suzuki Woodworking Shop•Yukio Suzuki
Matsuda Paulownia Box Workshop•Yukio Matsuda
Yamada Paulownia Box Workshop•Takao Yamada
Ohno Paper Ware Workshop•Masaji Ohno
Kasukabe city Paulownia Box Manufacturers' Cooperatives

Wood Package 木の箱

発行日……1987年5月10日
著者……広橋桂子
杉村敏男
U.G.サトー
岡田宏三
新井 亮
新井明子
秋月 繁
装幀・レイアウト……羽良多平吉+郷坪浩子 写真撮影……福田匡伸
翻訳……林 千根
写植……オギー有限会社
編集担当……久保田啓子
発行者……橋本周平
発行所……株式会社六耀社
東京都新宿区新宿2-19-12 静岡銀行ビル
〒160
TEL.03-354-4020(代)
FAX.03-352-3106
製版・印刷……日本写真印刷株式会社
製本……大口製本印刷株式会社

ISBN4-89737-060-4 C3070